Lecture Notes in Mathematics

Edited by A. Dold and B. Eckmann

1257

T0219705

Xiaolu Wang

On the C*-Algebras of Foliations in the Plane

Springer-Verlag

Berlin Heidelberg New York London Paris Tokyo

Author

Xiaolu WANG
Department of Mathematics
University of Chicago
Chicago, Illinois 60637, USA

Mathematics Subject Classification (1980): 34C35, 46L55, 46L80, 57R30

ISBN 3-540-17903-8 Springer-Verlag Berlin Heidelberg New York
ISBN 0-387-17903-8 Springer-Verlag New York Berlin Heidelberg

Printed in Germany

Printing and binding: Druckhaus Beltz, Hemsbach/Bergstr.
2146/3140-543210

TO MY FATHER

I would rather schematize the structure of mathematics by a complicated graph, where the vertices are the various parts of mathematics and the edges describe the connections between them. These connections sometimes go one way, sometimes both ways, and the vertices can act both as sources and sinks. The development of the individual topics is of course the life and blood of mathematics, but, in the same way as a graph is more than the union of its vertices, mathematics is much more than the sum of its parts. It is the presence of those numerous, sometimes unexpected edges, which makes mathematics a coherent body of knowledge, and testifies to its fundamental unity.

——— A. Borel

I have the feeling that we don't understand at all the extraordinary interplay of combinatorics and what I would call "conceptual" mathematics.

——— J. Dieudonné

Contents

§0. Introduction

0.1 In [C1] A. Connes introduced C*-algebras of foliations as an important ingredient in his noncommutative integration theory. Linking topology, geometry and analysis, it has become an important area of research; see [C2] for a more detailed treatment. Many general properties of C*-algebras of foliations have also been studied by T. Fack, G. Skandalis, M. Hilsum and others; see [F-S], [F], [H-S].

A very interesting and natural problem is to determine for various manifolds the structure of the C*-algebras of special types of foliations, or at least their K-theory. This will help us understand how the algebraic and topological invariants of the C*-algebras are related to the geometric and topological properties of foliations. For this aspect, see for instance [T], [N-T1], [N-T2], [N1], [N2] and [Tor].

So far the investigation has been carried out for only a few closed manifolds, i.e., the manifolds which are compact and without boundary. A closed manifold does not necessarily admit a nonsingular foliation. For instance, it is well known that the only closed connected orientable two-manifold which does is the torus. The C*-algebras of foliated tori have been studied by A. M. Torpe in [Tor]. In order to carry the investigation further, especially to explore the potential application of C*-algebras in dynamical systems, it is necessary to consider C*-algebras of foliated *open* manifolds, or what amounts to the same thing, to consider singular foliations of closed manifolds. The purpose of this paper is to initiate a study in this direction.

The most important kind of open two-manifolds are the punctured closed two-manifolds. Thus our study will be focused on foliations of (topological) two-manifolds with isolated singularities. Since all closed two manifolds have the plane as their universal covering space, except the two-sphere which is however the one-point compactification of the plane, the plane is the simplest yet most important open two manifold.

0.2 In this paper we discuss a method which will allow us to compute the "ordinary" foliations of the plane (Definition 1.5.12). A crucial conception involved is separatrices and canonical regions (1.5.1), which were studied in qualitative theory of ordinary differential equations. Roughly, a separatrix is a leaf whose arbitrary small neighborhoods contain leaves behaving differently from it in the large. The canonical regions in a foliation are just the complement regions of the separatrices. By definition, an *ordinary foliation* is topologically conjugate to a foliation in which the set of separatrices has planar measure zero. A foliation is called *with* T_2*-separatrices* if the set of separatrices is Hausdorff with the quotient topology in the leaf space. We give a complete classification of all the C*-algebras of foliations with T_2-separatrices up to isomorphism. This study shows how noncommutative "CW 1-complexes" appear naturally in C*-algebras. The "glueing" construction is used extensively. It suggests that the method of surgery in topology may be adopted in the noncommutative context. Our investigation also establishes a direct contact between C*-algebras, and the qualitative theory of ordinary differential equations. It is shown that the solution of the classification of C*-algebras of C^∞-foliations of the plane provides in the meantime the solution of the classification of the C*-algebras of all the 1-parameter continuous transformation groups of the plane (cf. §1.1). In fact C*-algebras of foliations are the prototypes of C*-algebras of transformation groups [E-H]. These two notions are elegantly unified in terms of C*-algebras of groupoids (cf. [Ren]). This approach is essential in our investigation (e.g. the proof of Theorem 4.1.2).

0.3 Our study is based on the remarkable work of Wilfred Kaplan [K1], [K2] and Lawrence Markus [M] which in turn is based on the previous works of I. Bendixon, H. Poincaré, H. Whiteney, E. George, E. Kamke and many others. The Kaplan-Markus theory provides a complete classification of the foliations of the plane up to topological conjugacy in terms of certain configurations known as Kaplan diagrams (see Ch. 11 [Bec]). Since nonconjugate foliation of the plane may very well have isomorphic C*-algebras, the Kaplan diagrams are not an invariant for C*-algebras.

Example. The two foliations illustrated by Fig. 0.1 are not topologically conjugate. Later (Example 4.3.3) we shall see that they have isomorphic C*-algebras.

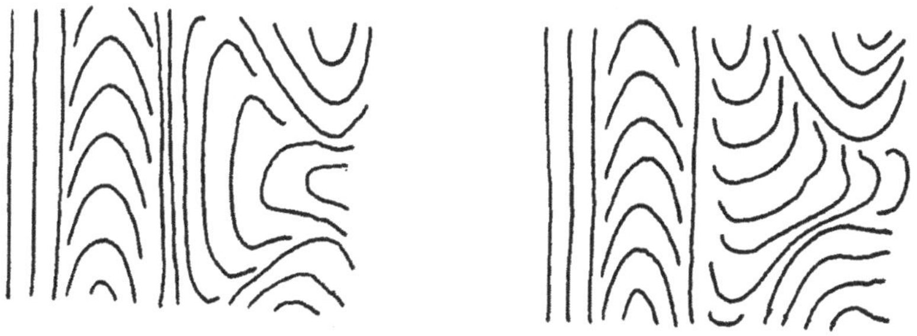

Fig. 0.1

Thus not only do we need to determine the C*-algebra of the foliation, represented by an arbitrary Kaplan diagram, but also to find a new configuration which is a complete intrinsic invariant for these C*-algebras. This invariant for all foliations with T_2-separatrices is what we shall call distinguished trees (Definition 3.3.1).

0.4 From the point of view of noncommutative topology, the way we study the C*-algebras of foliations is a natural analogue to that of studying topological manifolds by triangulation. We decompose a foliated manifold into saturated connected components. What equivalence relation among the leaves we choose to yield the decomposition may depend on the types of manifolds. For instance, one can use Novikov connected components for closed manifolds. For the plane region, our decomposition yields the standard canonical regions and separatrices defined for topological dynamical systems. This step is obvious when one recalls that the C*-algebra of a foliation is the reduced C*-algebra of the holonomy groupoid, which is the desingularization of the leaf space. In fact, such a decomposition divides the leaf space into "homogeneous" components.

We then compute the reduced C*-algebra of the restricted holonomy groupoid of each connected component along with its boundary. These C*-algebras are the "cells". A noncommutative "CW-complex" is then constructed by glueing these cells in a certain way. There are two types of glueing conditions in general: "local" and "global". The "local glueing" condition is fulfilled by taking the fibred product of the two "nearby" C*-algebras. The "global glueing" condition has, however, no analogy in the commutative context. It is fulfilled by imposing conditions on continuous fields of C*-algebras. The distinguished trees are a combinatorial-topological invariant. Their combinatorial structure contains the "local" information while their topological structure contains the "global" information.

In order to illustrate this method without involving too much combinatorial complicacy, we derive only the classification theorems for the C*-algebras of all foliations with T_2-separatrices. Our method may be used to obtain a possible classification of C*-algebras of all ordinary C*-algebras.

Finally, of course, we need to show that the "CW-complex" constructed in this way is indeed isomorphic to the C*-algebra of the foliation we begin with.

0.5 In §1, we start by clarifying some technical conditions concerning smoothness of foliations. Then we introduce the main results of Kaplan-Markus theory in a way convenient for our purpose. The main references are [Bec], [K1], [K2] and [M]. The central concept is that of the Kaplan diagram of a foliation (see 1.2.2), and its intrinsic definition (Definition 1.4.5). The fact that the Kaplan diagram of a foliation satisfies the axioms in Definition 1.4.5, together with Theorems 1.4.8 and 1.4.9 are the main results of the Kaplan-Markus theory.

The abstract formulation of Kaplan diagrams, i.e., "normal chordal systems", which is Kaplan's original approach, is postponed to 3.4.9, after we have clarified the relation between ℝ-trees and abstract chordal systems (Theorem 3.4.4). The classification theorem is reformulated in Theorem 3.4.10 (Theorem, p.11 [K2]).

Distinguished trees are, in particular, a class of $\mathbb{R}$-trees. In §2, we discuss briefly $\mathbb{R}$-trees as defined by John Morgan and Peter Shalen ((II.1.2) [M-S1]). By definition $\mathbb{R}$-trees are metric spaces in which any two points are joined by a unique arc with the distance given by the arc length. As generalized trees, thev had been studied in [A-M]. Our emphasis is on regular $\mathbb{R}$-trees (Definition 2.3.3) which are the separable $\mathbb{R}$-trees equipped with "vertices" and "edges". They can be regarded as the "minimal" generalization of trees (see Theorem 2.3.5). We generalize the construction of quotient trees to the context of $\mathbb{R}$-trees in 2.4. In order to study Kaplan-Markus flows in multi-connected regions in a later paper, we introduce in 2.5 the concept of $\mathbb{R}$-graphs in passing.

The essential results of this paper are contained in §3 and §4. Distinguished trees are characterized in §3. They are a special class of regular $\mathbb{R}$-trees in which the vertices and edges are $\mathbb{Z}_2$-graded in a certain way (Definition 3.3.1). We first show how to associate an $\mathbb{R}$-tree to a foliation with T_2-separatrices (Definition 3.2.13). These trees are shown to be distinguished trees (Theorem 3.3.4). Conversely, every distinguished tree is shown to be a tree associated to a foliation with T_2-separatrices (Theorem 3.5.12). The special case where the $\mathbb{R}$-trees are actually trees is given an independent proof (Theorem 3.3.7).

Finally we come back to C*-algebras of foliations in §4. Recall the program laid out in 0.4. There are in general five types of canonical regions which have been given in Corollary 1.5.8. The C*-algebras of parallel foliations of polygonoids are determined explicitly in Theorem 4.1.2. In analogy with the construction of "a tree of groups" in combinatorial group theory, for every distinguished tree T we associate a tree of C*-algebras (T,A,σ) (Definition 4.2.9). In analogy with amalgamated products of trees of groups, here we define for a system (T,A,σ) a C*-algebra $C^*(T,A,\sigma)$, called the global fibred product, which is independent of a particular choice of the set σ of homomorphisms, up to isomorphisms of C*-algebras (Theorem 4.2.19).

Thus we may write $C^*(T)$ for $C^*(T,A,\sigma)$. The correspondence between the distinguished trees and the C*-algebras has the following property (Theorem 4.2.21).

Theorem. *Two distinguished trees* T_1 *and* T_2 *are similar iff the* C*-*algebras* $C^*(T_1)$ *and* $C^*(T_2)$ *are isomorphic.*

Then we show that the C*-algebra of a foliation (Ω,F) of the plane is isomorphic to $C^*(T(F),A,\sigma)$ for some σ, where $T(F)$ is the distinguished tree of the foliation (Ω,F) (Theorem 4.3.1). Combining with the results obtained in §3, we conclude (Theorem 4.3.2)

Theorem. *The distinguished trees are a complete isomorphism invariant for the* C*-*algebras of foliations with* T_2-*separatrices.*

0.6 Note that a foliation of the plane is equivalent to a foliation of the two-sphere with a singular point of arbitrary type. Our study can be carried over to flows on the two sphere with isolated singularities. "Blowing up" these singular points, we pass over to the situation of Kaplan-Markus flows on multiconnected open regions. Instead of regular $\mathbb{R}$-trees and distinguished trees, we shall have regular $\mathbb{R}$-graphs and distinguished graphs. This will be discussed elsewhere.

0.7 $\mathbb{R}$-trees can be considered to be degenerate hyperbolic spaces. They play an important role in Morgan and Shalen's remarkable work [M-S1] and [M-S2]. There is a close connection between the construction of an $\mathbb{R}$-tree for a measured lamination on a closed surface and the construction of the $\mathbb{R}$-tree $T(F)$ for a foliation of the plane described in §3. In fact, let M be a closed surface with a measured lamination F, $\chi(M) \leq 0$. Then F can be lifted to a measured lamination $\tilde{F}$ on the plane Ω. The complement of $\tilde{F}$ in Ω is a union of open regions, each of which admits parallel foliations. If jointly this gives a foliation of Ω (this is always the case if we can choose the leaves of the foliation of each open region "parallel" to its boundary), then the construction in §3 defines an $\mathbb{R}$-tree $T(\tilde{F})$, which is homeomorphic to the $\mathbb{R}$-tree defined by the measured lamination. There are also interesting relations between the C*-algebra of a singular foliation of a closed surface and the C*-algebra of its singular covering foliation of the plane. The $\mathbb{R}$-graphs corresponding to these two C*-algebras are related through the actions of the fundamental group of the surface (cf. §4.1). We will carry out the investigation of this relation elsewhere.

0.8 Conventions

Throughout this paper unless otherwise stated, we use the open unit disc Ω in $\mathbb{R}^2$ as the topological model for the plane. We denote by $\partial\Omega$ the unit circle in $\mathbb{R}^2$. For any metric space X, $x \in X$, $\delta > 0$, we write $B(x,\delta)$, the open ball of radius δ and centered at x.

For a saturated submanifold V of (Ω,F) (with or without boundary), we shall denote by $\mathbf{G}(V,F)$ the graph of the restricted foliation of V, and by $C^*(V,F)$ the corresponding C^*-algebra (cf. the beginning of 4.1). By an ideal of a C^*-algebra, we always mean a closed two-sided ideal.

The bounded operators and compact operators in a Hilbert space H will be denoted L(H) and K(H) respectively.

The author is grateful to Professor Marc A. Rieffel for his invaluable suggestions and extensive comments during the preparation of this manuscript. The author also wishes to express his gratitude to Professor Adrian Ocneanu for a helpful discussion at an earlier stage and to Professor Jonathan Rosenberg for suggesting a simplified proof of Theorem 4.1.2. Finally, he wishes to thank Professors R. Hermann, Morris Hirsch and Charles Pugh for providing the literature about foliations.

§1. Foliations of the plane

A complete classification of foliations of the plane up to topological conjugacy was carried out by Wilfred Kaplan in [W1] and [W2] following the earlier works of I. Bendixon, E. Kamke, H. Poincaré, H. Whiteney and many others ([Ben], [Kam], [Po], [Wh]). Although the main results of Kaplan's theory are intuitively convincing and quite elementary in appearance, the proofs of this kind of classification theory are very complicated since as a general rule the combinatorial analysis involved is messy.

Kaplan's theory grew out of the qualitative theory of ordinary differential equations which in fact provides its most important and direct application. This aspect was studied by L. Markus from the viewpoint of the geometry of the global structure of solutions of ordinary differential equations, cf. [M].

Our aim in this section is to give a concise description of the main results of Kaplan-Markus theory so the reader may avoid getting into all the technical details of Kaplan's original work. Also we shall fix some notations along the way for the following sections, and this makes our exposition self-contained to a large extent.

We shall discuss some basic properties of foliations of the plane in 1.1 and clarify the points involving orientability of foliations and the possibility of smoothing a C^0-foliation to a C^∞-foliation. The need for smoothness arises from the study of C*-algebras of foliations, which is our main concern in this paper.

The presentation of Kaplan-Markus theory in 1.2-1.4 basically follows the approach of A. Beck (Ch. 11 [Bec]). The proofs of most theorems are omitted but appropriate references are given. The main result is the classification of foliations of the plane up to conjugacy in terms of the Kaplan diagrams.

In 1.2 we describe how to construct a Kaplan diagram for a given foliation (Ω,F). In 1.3 we shall illustrate by examples what sorts of foliations and Kaplan diagrams can occur. Then one can guess what should be an axiomatic definition of a Kaplan diagram and how to recover a foliation from a given diagram. This is discussed in 1.4. The reader may then draw a few pictures by himself and play the game both ways.

1.1 Orientability and smoothness of foliations

First we recall some definitions.

Definition 1.1.1. A C^k-foliation (M,F), $1 \leq k \leq \infty$, is *orientable* if the tangent bundle TF of the foliation is an orientable vector bundle. The foliation (M,F) is *transversally orientable* if the normal bundle of the foliation, TM/TF, is orientable.

Proposition 1.1.2. *Assume that* M *is simply connected. Then any* C^k*-foliation* (M,F), $1 \leq k \leq \infty$, *is orientable and transversally orientable.*

Proof. This follows from the fact that any real vector bundle over a simply connected space is orientable. Q.E.D.

Proposition 1.1.3. *If* M *is an orientable manifold then a* C^k*-foliation* (M,F), $1 \leq k \leq \infty$, *is orientable iff it is transversally orientable.*

Proof. Since TM is orientable, this follows from the "two out of three" property of orientability for vector bundles. Q.E.D.

For a C^0-foliation (M,F), in general the tangent bundle TF does not exist. If the leaves of F are one-dimensional, we shall say that (M,F) is orientable if it arises from a C^0-transformation group $(M,\mathbb{R})$, or equivalently, arises from a C^0-flow. See [E-H] and [Bec] for definitions of transformation groups and flows.

1.1.4. It follows from Proposition 1.1.2 that any C^1-foliation of the plane is orientable and transversally orientable. A less trivial fact is that any C^0-foliation of the plane is orientable (Corollary to Theorem 42, [Kl]). An orientable C^k-foliation of codimension 1 of the plane is given by a C^k-flow, equivalently, by a C^k-action of the reals $(0 \leq k \leq \infty)$. By means of this connection, Connes' theory of C^*-algebras of foliation provides us a very powerful tool for handling the more classical problems of classifying 1-parameter transformation group C^*-algebras. However C^*-algebras are defined so far only for $C^{\infty,0}$-foliations, i.e., for foliated manifolds which are C^∞ in the leaf direction but only C^0 in the transversal direction, and we would like to classify the C^*-algebras of all the $\mathbb{R}$-transformations of C^0-classes on the plane. This point is clarified again by a theorem of W. Kaplan. First we recall a definition.

Definition 1.1.5. Let (M,F) and (N,F) be foliated manifolds. We say that (M,F') and (N,F') are *topologically conjugate* or just *conjugate* if there is a homeomorphism from M onto N which maps leaves onto leaves.

Theorem 1.1.6 (Thm 1 [K3]). *Let* F *be a* C^0*-foliation of the plane* $\mathbb{R}^2$. *Then* $(\mathbb{R}^2, F)$ *is topologically conjugate to a foliation of the disc* (Ω, F') *such that the leaves of* F' *are the level curves of a harmonic function in* Ω. *In particular* (Ω, F') *is a* C^∞*-foliation.*

For flows (with isolated singularities) on 2-manifolds similar assertions are "usually" true. This was shown recently by C. Gutierrez ([Gut]) based on results of Denjoy, A. Schwartz and others. Roughly speaking, C^0-flows can always be smoothed to C^1, while C^2-flows can always be smoothed to C^∞. The only obstruction for smoothing a C^1-flow to a C^2-flow is the existence of *exceptional components* of foliations of the torus, which are obtained by suspending Denjoy homomorphisms of the circle. More precisely, a C^0-foliation F of a two manifold M is topologically conjugate to a C^∞-foliation (M, F') iff (M, F) does not contain exceptional components of foliations. We will discuss this in more detail in a separate paper.

Let (M, F) be any C^0-foliation. If (M, F) is smoothable, i.e., topologically conjugate to a C^∞-foliation (M', F'), then we may regard the C*-algebra of the foliation (M, F) to be $C^*(M, F')$. Suppose (M, F) is topologically conjugate to another C^∞-foliation (M, F''). Then (M, F') is topologically conjugate to (M, F'').

Proposition 1.1.7. *Let* (M, F) *and* (M', F') *be two topologically conjugate* $C^{\infty,0}$*-foliations. Then the two C*-algebras* $C^*(M, F)$ *and* $C^*(M', F')$ *are isomorphic. In particular up to isomorphism, the C*-algebra* $C^*(M, F)$ *does not depend on the differential structure of* (M, F').

Proof. Let the dimension and codimension of (M,F) be p and q respectively. Let T be a C^0-transversal of (M,F) of dim q, in the sense that every point $x \in T$ has a fundamental neighborhood (U_x, ϕ_x) with $\phi_x(U_x) = D_1^p \times D_1^q$, and $\phi_x(U_x \cap T) = \{0\} \times D_1^q$. Assume that T is faithful, i.e., T intersects every leaf of (M,F) (p. 202 [H-S]). Here the coordinate system $\{(U_x, \phi_x)\}$ is of $C^{\infty,0}$-class. Let h be a homeomorphism conjugating (M,F) and (M',F'). Then $T' = h(T)$ is a faithful C^0-transversal of (M',F'). It is easy to see that T (T', resp.) is a subset of G_T^T ($G_{T'}^{T'}$, resp.) (see §1, Part I for notation), both open and closed (p.6 [Ren]). Thus both G_T^T and $G_{T'}^{T'}$ are r-discrete groupoids (Definition 1.2.6 [Ren]). The conjugacy h between (M,F) and (M',F') induces a bijection between G_T^T and $G_{T'}^{T'}$ in an obvious way. By (Lemma 1.2.7 [Ren]), the Haar system on the two groupoids is essentially the counting measure system, hence equivalent. Thus the two topological groupoids G_T^T and $G_{T'}^{T'}$ are isomorphic. It follows that $C^*(G_T^T)$ and $C^*(G_{T'}^{T'})$ are isomorphic, and the Hilsum-Skandalis stability theorem ([H-S]) implies that $C^*(M,F) \simeq C^*(M',F')$. Q.E.D.

Alternatively, when M is the unit disc Ω, we may define the C^*-algebra of a C^0-foliation (Ω,F) to be the C^*-algebra of a C^0-transformation group $(\Omega,\mathbb{R})$ which produces the foliation. Such a transformation groups exists by 1.1.4 and is clearly also unique up to conjugacy of transformation groups [E-H]. It is an easy fact that a pair of conjugate transformation groups have isomorphic C^*-algebras. If an $\mathbb{R}$-transformation group over the disc is C^∞, which produces the foliation (Ω,F), then the graph G of (Ω,F) is homomorphic to $\Omega \times \mathbb{R} \simeq \mathbb{R}^3$ and we have $C^*(\Omega,F) \simeq C_0(\Omega) \times \mathbb{R}$ by Proposition 1.4, Part I.

1.1.8 Summarizing the discussion in 1.1.7, we conclude that to classify continuous $\mathbb{R}$-transformation groups over the plane, it suffices to consider C^∞-foliations.

1.2 Kaplan diagrams of foliations

1.2.1 Let (Ω,F) be a foliation of the unit disc Ω. We first observe that (Ω,F) does not contain closed leaves. In fact a closed leaf would bound a compact disc which has Euler characteristic 1. Since the restricted foliation on this compact disc is orientable (1.1.2) the Poincaré-Hopf theorem asserts that there is at least one singularity contained in the flow, cf. Thm 13 [K1]. Thus every leaf L is open and $\bar{L}\setminus L \subset \partial\bar{\Omega}$. Moreover we may assume that $\bar{L}\setminus L$ consists of exactly two ends of L (cf. Thm 29 [K2]).

Proposition 1.2.1a. *Every leaf* L *in the leaf space* Ω/F *is closed with respect to the quotient topology.*

Proof. Every leaf L is a closed subset of Ω ([K1] Thm 16). Q.E.D.

1.2.2 Foliations of the open disc Ω are classified up to conjugacy by the Kaplan diagrams. A Kaplan diagram is a certain system of chords in $\bar{\Omega}$, called a "normal chordal system" by Kaplan (p. 173 [K1]). We shall need Kaplan's formulation of normal chordal systems in §3.4 (see Definition 3.4.7). Here we opt for a more explicit approach of Beck (Ch. 11 [Bec]).

The chords in a Kaplan diagram are in one-to-one correspondence with the leaves of the foliation (Ω,F). Let us now describe the construction. Every C^0-foliation is topologically conjugate to a C^∞-foliation with leaves of planar measure zero (cf. p.354 [Bec]). By Proposition 1.1.7, we shall assume this is the case. Fix an arbitrary point $x_0 \in \Omega$. Let L_{x_0} be the leaf containing x_0. By the Jordan curve theorem, L_{x_0} divides Ω into two regions D_{x_0} and D'_{x_0} with areas $m(x_0)$ and $(\pi - m(x_0))$ respectively. Fix a chord $C(x_0)$ in $\bar{\Omega}$ which divides $\partial\Omega$ into two arcs with lengths $\ell(x_0)$ and $(2\pi - \ell(x_0))$ such that $m(x_0)/(\pi - m(x_0)) = \ell(x_0)/(2\pi - \ell(x_0))$. It follows that $\ell(x_0) = 2m(x_0)$ as scalars. For every point $x \notin L_{x_0}$ we shall give a procedure to choose some chord $C(x)$ which again divides $\partial\Omega$ in the same ratio as L_x divides the area of Ω. It turns out that we can pack all such chords in $\bar{\Omega}$ in such a way that they form a disjoint maximal system of chords satisfying certain special conditions.

The leaf L_x also divides Ω into two regions D_x and D'_x. They are labelled so that $D_x \cap D_{x_0} \neq \emptyset$. Thus $D_x \cap D_{x_0}$ is either D_x or D_{x_0}; see Fig. 1.1.

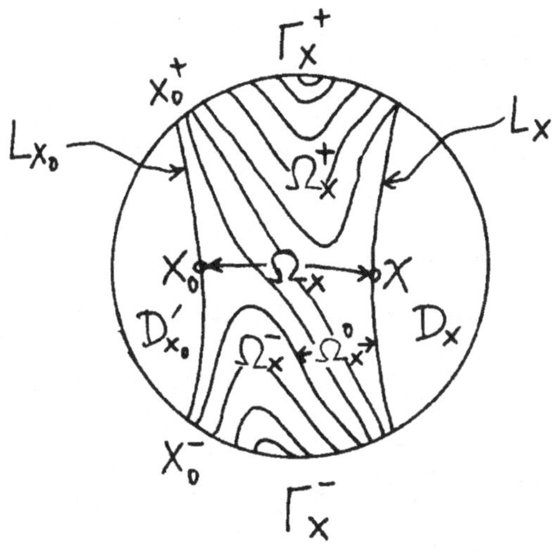

Fig. 1.1

Assume first that $D_x \cap D_{x_0} = D_x$. The region between L_{x_0} and L_x is then $D_{x_0} \cap D'_x$, which we denote by Ω_x. Let x_0^+ and x_0^- be the two ends of L_{x_0} on $\partial\Omega$. Let the two arcs consisting of $\partial\Omega \cap \bar{\Omega}_x$ be Γ_x^+ and Γ_x^-, which contain x_0^+ and x_0^-, respectively. Let

$$\Omega_x^+ = \{y \in \Omega_x | L_y \text{ has both ends on } \Gamma_x^+\}$$

$$\Omega_x^0 = \{y \in \Omega_x | L_y \text{ has one end in } \Gamma_x^+, \text{ the other in } \Gamma_x^-\} ;$$

similarly for Ω_x^-. Thus $\Omega_x = \Omega_x^+ \cup \Omega_x^- \cup \Omega_x^0$ is a disjoint union.

Let the planar measures of D_x, Ω_x^+, Ω_x^- and Ω_x^0 be $m(x)$, $m^+(x)$, $m^-(x)$ and $m^0(x)$ respectively. Then

$$m(x_0) = m^+(x) + m^-(x) + m^0(x) + m(x) . \tag{1.1}$$

Recall that the chord $C(x_0)$ cuts off an arc of length $2m(x_0)$ from $\partial\Omega$. We mark off at the corresponding ends of this arc the arcs of lengths $2m^+(x) + m^0(x)$ and $2m^-(x) + m^0(x)$. From (1.1) this leaves an arc of length $2m(x)$ and we denote the corresponding chord by $C(x)$.

A similar construction applies to the case where $D_{x_0} \cap D_x = D_{x_0}$.

Lemma 1.2.3. *Assume that the construction above yields a chord* $C(y)$ *for another point* $y \in D(x_0) \setminus L_x$. *Then* $C(x)$ *and* $C(y)$ *do not intersect in* Ω.

Proof. Case i. Assume that $x \in D(y)$. Then $\Omega_y^+ \subseteq \Omega_x^+$, $\Omega_y^- \subseteq \Omega_x^-$ and $\Omega_y^0 \subseteq \Omega_x^0$. It follows that $2m^+(y) + m^0(y) \leq 2m^+(x) + m^0(x)$ and $2m^-(y) + m^0(y) \leq 2m^-(x) + m^0(x)$. Moreover equality cannot be true simultaneously. Otherwise, adding them up side by side, we get $2m(\Omega_y) = 2m(\Omega_x)$, a contradiction.

Case ii. Assume $y \in D(x)$. The result is the same.

Case iii. Neither $x \in D(y)$ nor $y \in D(x)$. In Kaplan's terms, x_0, x, y form a cyclic set. Without loss of generality, assume that $y \in \Omega_x^+$. Then $D(y) \cup \Omega_y^+ \subset \Omega_x^+$ and $D(x) \cup \Omega_x^- \subset \Omega_y^-$ (see Fig. 1.2). If $m^0(x) \geq m^0(y)$ then $2m^+(x) + m^0(x) \geq 2m^+(y) + 2m(y) + m^0(y)$. If $m^0(x) \leq m^0(y)$ then $2m^-(y) + m^0(y) \geq 2m^-(x) + 2m(x) + m^0(x)$. In either case $L(x)$ and $L(y)$ can not meet except possibly on $\partial\Omega$.

Similarly, if $y \in \Omega_x^-$. Q.E.D.

1.2.4. The chordal system $\{L(x)\}$ obtained in this way is independent of the initial choice of the base point x_0 up to similarity (cf. 1.4.7) and it is called the Kaplan diagram of the foliation (Ω, F), denoted by $K(F)$.

The transformation sending leaves L_x of (Ω, F) to chords $L(x)$ in $K(F)$ will be denoted Δ and called the Kaplan map.

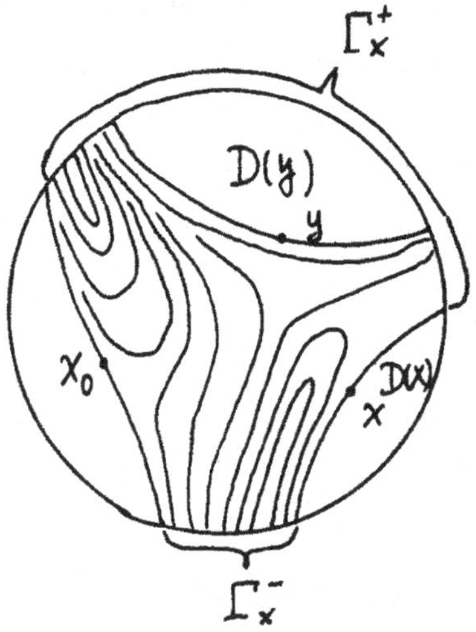

Fig. 1.2

1.3. **Examples of foliations and their Kaplan diagrams**

1.3.1 The foliation of $\mathbb{R}^2$ with one Reeb component is given in Fig. 1.3(a). The corresponding foliation of Ω and its Kaplan diagram are given in Figs. 1.3(b) and (c), respectively.

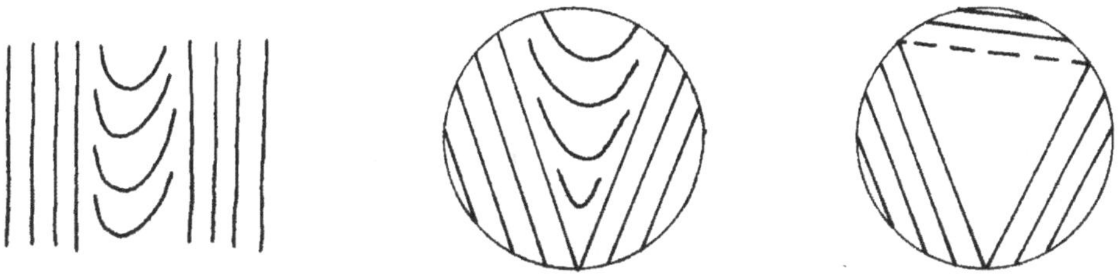

(a) (b) (c)

Fig. 1.3

We note that the foliations of the plane illustrated in Fig. 1.4 and Fig. 1.3(a) are conjugate. The foliation described in Fig. 1.5 is conjugate to the constant foliation of the plane (although one obtains the familiar orientation preserving Reeb component of the annulus by taking its quotient). The reader should also convince himself that the foliation of the lower half plane shown in Fig. 1.6 is also conjugate to the constant foliation of the plane.

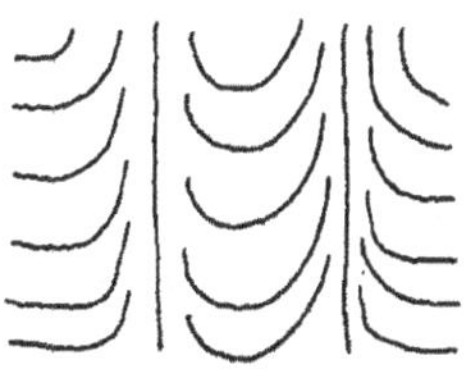

Fig. 1.4

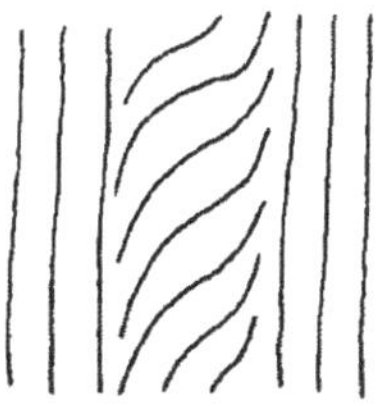

Fig. 1.5

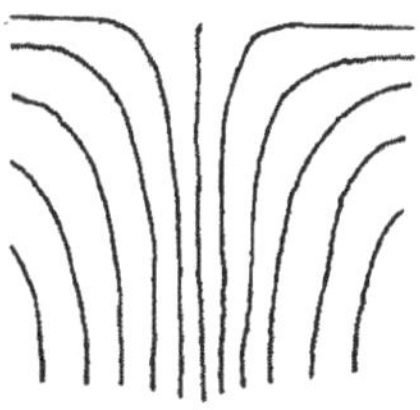

Fig. 1.6

1.3.2 If we consider the Reeb component in Fig. 1.3(a) as a "river" with two "banks", then in Fig. 0.1(a) or (b), one finds a "branched river" with three "banks". Later we shall see that the "source" corresponds to the *open side* of the corresponding *polygonoid* in the Kaplan diagram. The Kaplan diagrams of the foliations shown in Fig. 0.1(a) and (b) are given in Fig. 1.7(a) and (b).

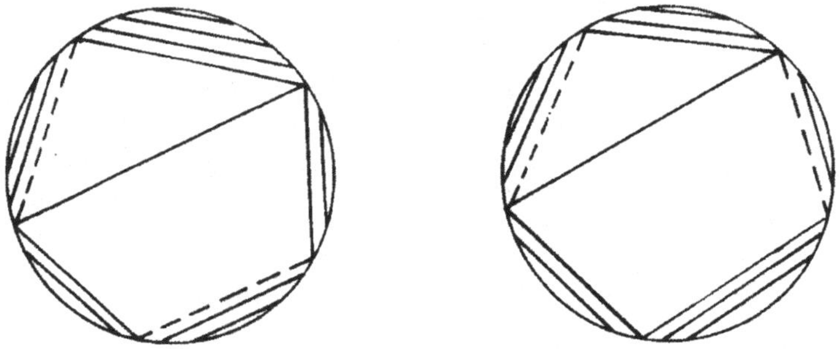

(a) (b)

Fig. 1.7

1.3.3 The above examples (Fig. 1.3) can be generalized to "multi-branched rivers", even with a countably infinite number of "banks". Here are two examples which have been put on the unit disc:

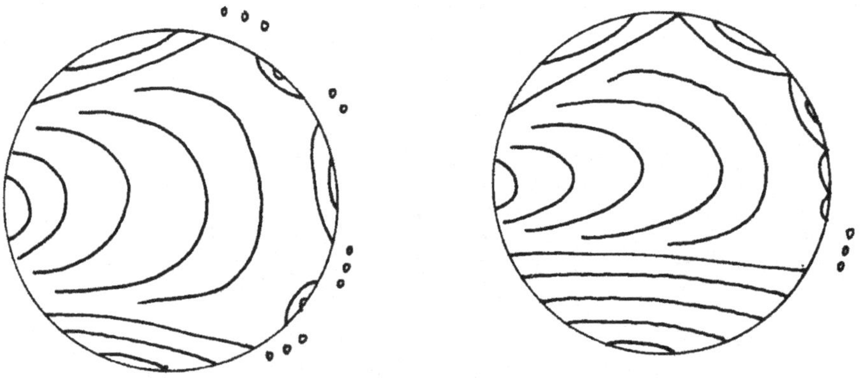

Fig. 1.8

The corresponding Kaplan diagrams are:

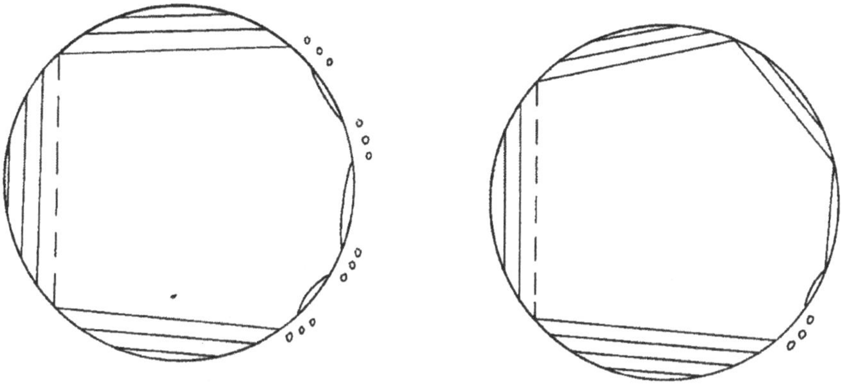

Fig. 1.9

Note that in Fig. 1.9(a) after taking away these open arcs which correspond to the chords, the remaining part of $\partial\Omega$ is a Cantor set of measure zero.

1.3.4 We may build foliations of the plane by combining countably many of the types shown in previous examples. We give two Kaplan diagrams in Fig. 1.10 and Fig. 1.11 and it is easy to see what foliations they represent.

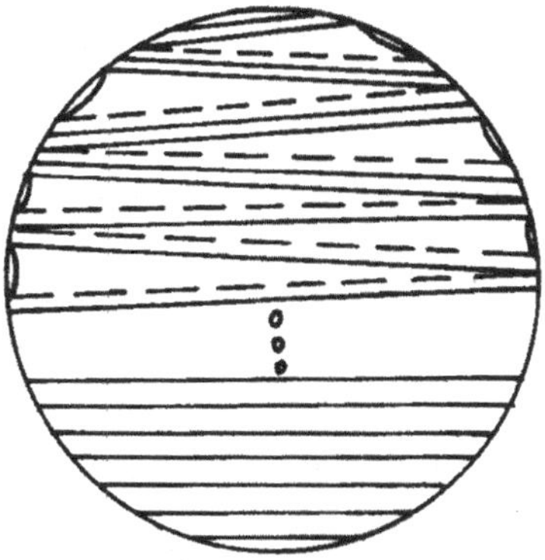

Fig. 1.10

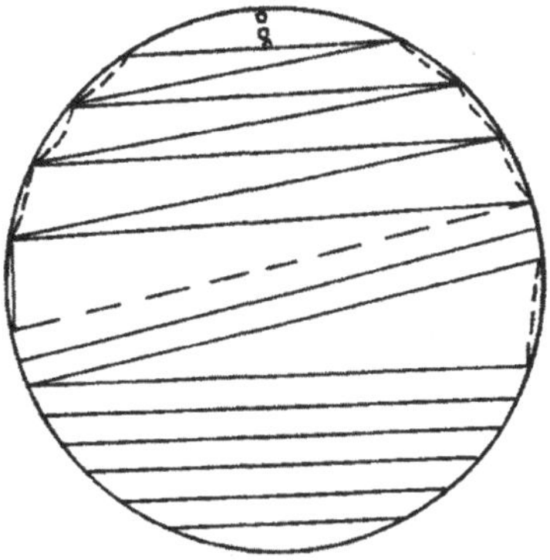

Fig. 1.11

1.3.5 The curve family filling the disc Ω shown in Fig. 1.12(a) is *not* a foliation. It is easy to see that the point 0 has no fundamental neighborhood. Thus the corresponding chordal system in Fig. 1.12(b) is not a Kaplan diagram.

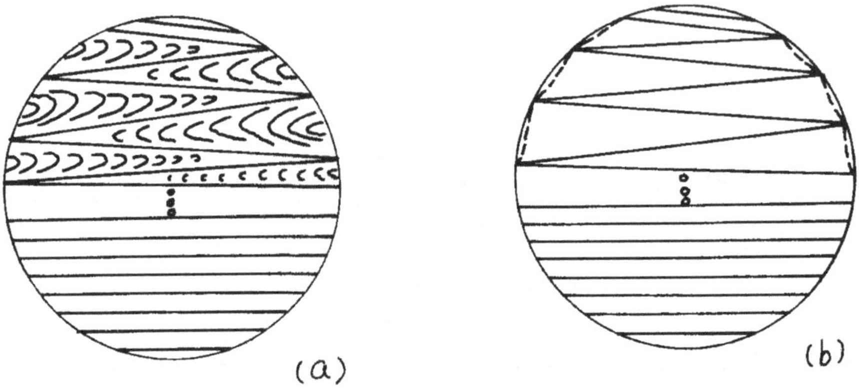

Fig. 1.12

1.4. Classification of foliations of the plane

1.4.1 Let $K(F)$ be the Kaplan diagram of a foliation (Ω, F). If (Ω, F) is not conjugate to the trivial foliation by parallel lines, then the chords in $K(F)$ do not fill up the disc $\bar{\Omega}$. The complement of all these chords in general is a countable disjoint union of convex figures. Each convex figure along with its boundary is a "polygonoid". Those polygonoids play a crucial role in the Kaplan diagram. We have to make this more precise before we give an intrinsic definition of Kaplan diagrams.

Definition 1.4.2. A *polygonoid* P *in* $\bar{\Omega}$ is a closed convex region in $\bar{\Omega}$ such that its boundary ∂P consists of countably many closed chords

$S_1, S_2, \ldots$ together with a measure zero closed subset F of $\partial\Omega$, where the S_i's do not intersect each other except possibly at the ends and all ends of S_i's are contained in F (e.g. Fig. 1.8(a)). The chords S_i are called the *sides* of the polygonoid P.

When a polygonoid has only finitely many sides, it degenerates to a polygon and the measure zero closed subset is just a finite subset of $\partial\Omega$, namely, the vertices of the polygon.

1.4.3 Let $K(F)$ be the Kaplan diagram of a foliation (Ω, F). Let P be a polygonoid specified by $K(F)$ as in 1.4.1. Then all sides of P except exactly one are chords contained in $K(F)$ (p.357 [Bec]). In other words, all sides but exactly one correspond to certain leaves of the foliation (Ω, F) under the Kaplan transformation Δ defined in 1.2.3. The exceptional side which is not a chord in $K(F)$ is called the *open side* of the polygonoid P and the rest of the sides are the *closed sides* of P. The open side corresponds to the "source" of the "river".

We call a polygonoid *obtuse* if its open side is either the longest side or the second longest side among the sides contained in its boundary (see the triangles in Fig. 1.10). Otherwise P is called *acute* (see Fig. 1.12)

If a closed side of a polygonoid and an arc in $\partial\Omega$ bounds a region filled up by chords in $K(F)$, this region is called an *abutment* (see Fig. 1.9).

For a collection C of chords, we also let C represent the union of all the chords in C as a topological subset of $\bar{\Omega}$. What it stands for will be clear from the context.

Consider the examples 1.3.4 and 1.3.5. We see that most polygonoids in a Kaplan diagram are long and thin. Moreover, an infinite sequence of acute polygonoids has to converge to the boundary $\partial\Omega$.

Theorem 1.4.4. *Let* $K(F)$ *be a Kaplan diagram of a foliation* (Ω, F). *Let* $\{P_i\}$ *be an infinite sequence of distinct acute polygonoids of* $K(F)$. *Then the diameters of the polygonoids,* $\operatorname{diam}(P_i)$, *go to* 0.

Proof. See Theorem 11.10, p.361 [Bec]. Q.E.D.

At this point, we have stated all the essential properties of Kaplan diagrams, so now we can summarize and give an intrinsic definition.

Definition 1.4.5. A *Kaplan diagram* $K = (C, P)$ is a collection C of closed chords and a collection P of polygonoids in the closed unit disc $\bar{\Omega}$ (with Euclidean metric) such that:

(1) $C \cup \bigcup_{P \in P} P = \bar{\Omega}$

(2) For each $P \in P$, there is in $\bar{\Omega}$ one chord $s \notin C$ such that $P \cap C$ together with s is the union of all sides of P. (Then the chord s is called the *open side* of P and the chords in $P \cap C$ are called the *closed sides* of P.)

(3) If two distinct polygonoids P and P' in P have nonempty intersection, then $P \cap P'$ is a chord in C. (Then P and P' are said to be *adjacent*.)

(4) If two distinct chords C and C' in C have nonempty intersection, then $C \cap C'$ is a point in $\partial\Omega$ and C, C' are closed sides of either a polygonoid $P \in P$ or of two adjacent P, $P' \in P$.

(5) If $\{P_i\}$ is an infinite sequence of distinct acute polygonoids in P, then $\dim P_i \to 0$.

1.4.6 In Definition 1.4.5, Axioms (1) and (2) imply that the chordal system C has the following maximal property: If a chord s in $\bar{\Omega}$ fails to intersect the interior of any chords in C, then s is either the open side or a diagonal chord of some polygonoid $P \in P$ (cf. 11.8 [Bec]).

For simplicity, we say that a chord s is in K if s is in C. Let $K = (C,P)$; similarly for polygonoids.

1.4.7 Let K_1 and K_2 be two Kaplan diagrams in Ω. Suppose that there is a homeomorphism h of $\partial\Omega$ onto itself with the property that for any $x_1, x_2 \in \partial\Omega$, x_1 and x_2 are the ends of a chord in K_1 if and only if $h(x_1)$ and $h(x_2)$ are the ends of a chord in K_2. Then we say that the Kaplan diagrams K_1 and K_2 are *similar* and we call h a *similarity* between them. It is clear that when h is a similarity from $K_1 = (C_1,P_1)$ to $K_2 = (C_2,P_2)$, then h induces a bijection between P_1 and P_2.

Theorem 1.4.8. *Assume that* $K = (C,P)$ *is a Kaplan diagram as defined in 1.4.5. Then there is a foliation* (Ω,F) *such that the Kaplan diagram* $K(F)$ *of* (Ω,F) *as constructed in 1.2.2 is similar to* K.

Theorem 1.4.9. *Let* (Ω,F_1) *and* (Ω,F_2) *be two foliations of the disc* Ω. *Then the corresponding two Kaplan diagrams* $K(F_1)$ *and* $K(F_2)$ *are similar if and only if the two foliations* (Ω,F_1) *and* (Ω,F_2) *are conjugate.*

Theorems 1.4.8 and 1.4.9 are the main results in Kaplan's theory about classification of foliations of the plane. For the proofs, see Theorems 11.7, 11.10, 11.12 [Bec] and recall 1.1.4.

1.4.10 We observe that any triplet L_1, L_2, L_3 of distinct leaves of a foliation F satisfy exactly one of the five relations: $L_1|L_2|L_3$ (L_2 separates L_1 and L_3), $L_1|L_3|L_2$, $L_2|L_1|L_3$, $|L_1L_2L_3|^+$ (L_1, L_2, L_3 form a counterclockwise cyclic triplet), and $|L_1L_3L_2|^+$. The corresponding chords $\Delta(L_1)$, $\Delta(L_2)$, $\Delta(L_3)$ in C clearly also satisfy the same relation (1.2.4). Here C is the chordal system of the Kaplan diagram $K(F) = (C,P)$ of the foliation F.

It turns out that the abstract chordal system C equipped only with such a set of relations determines (Ω,F) up to conjugacy, and determines therefore $K(F)$ up to similarity. We observe that these chordal relations can be represented by the order relations of the endpoints of these chords. The concept of Kaplan diagrams which we adopt in this section is actually the geometric explicit realization of these abstract chordal systems which were called the "normal chordal systems" by Kaplan ([K1], [K2]). We shall discuss Kaplan's original construction in §3.4 and use it in §3.5.

1.5. Canonical regions and separatrices

A basic technique in studying the structure of a foliation of a manifold is to decompose the manifold into "components" or "regions", in which the foliations are homogeneous in a sense.

For foliations of the plane, the concept is provided by "canonical regions" and "separatrices". Both are well-defined in the geometric theory of differential equations. We shall discuss how these relate to the polygonoids and sides of the Kaplan diagram of the foliation.

1.5.1 In 1.2.2, we have seen that because of the Jordan curve theorem, every leaf L separates Ω into two distinct regions and two distinct leaves L_1, L_2 bound a unique region. We now define a topology on the leaf space Ω/F by decreeing that the leaves filling up the region bounded by two leaves is an element of a basis for this topology. Let S_0 be the set of leaves which are not closed in this topology. A leaf L contained in the closure of S_0 is called a *separatrix* of (Ω,F). Note that S_0 is in general not closed. A leaf L contained in the closure of $S_0\setminus\{L\}$ is called a *limit separatrix* of (Ω,F).

This topology on Ω/F shall be called the *coarse leaf topology*; it is in general coarser than the usual quotient topology on the leaf space, where any open set which is a union of leaves is open. Consider the example in 1.3.1. There are two separatrices. The coarse closure of each contains exactly itself and the other separatrices. However, every leaf is closed in the usual leaf topology.

1.5.2 The set of separatrices of (Ω,F) is closed in $(\Omega/F$, coarse topology$)$ by definition, thus is closed in $(\Omega/F$, quotient topology$)$. So the union of all separatrices is a closed subset of Ω. A connected component of the complement of the union of separatrices in Ω is called a *canonical region*. Two distinct canonical regions 0 and $0'$ are *adjacent* if there is a separatrix L such that $0\cup L\cup 0'$ is a connected region.

Warning. Although both the notions of separatrices and canonical regions can be well defined on multi-connected regions, they depend on the foliated manifold involved. It may very well happen that a separatrix in a foliated manifold will no longer be a separatrix when viewed as a leaf in a foliated submanifold. For example in Fig. 1.3(b) consider removing one of the separatrices and its abutment.

Lemma 1.5.3. *Let* O *be a canonical region of a foliation* (Ω,F). *Then* (O,F) *is conjugate to* Ω *with the trivial constant foliation.*

Proof. See Theorems 1, 2, 3 in [Wh]. Q.E.D.

1.5.4 **Warning.** The converse of Lemma 1.5.3 is not always true. In other words, it often happens that the union of many canonical regions along with the separatrices between them is conjugate to (Ω,triv). For example, see Fig. 1.3(b). In fact, roughly, decomposing (Ω,F) into such maximal components is what Kaplan called *normal subdivisions* (see Definition 3.4.9).

Lemma 1.5.5. *If* L *is a leaf in* (Ω,F) *such that* $\Delta(L)$ *(see 1.2.4) is a closed side of some polygonoid* P *of* $K(F)$, *then* L *is a separatrix.*

Proof. From the construction of the chordal system in 1.2 and Lemma 1.5.3, the closure of the one point set $\{L\}$ in the coarse leaf topology consists of all the leaves $\{L_i\}$ such that $\{\Delta(L_i)\}$ are precisely all the closed sides of P or of P, and P' if L is a side of two polygonoids P and P'. Since $\mathrm{Card}\{L_i\} > 1$, the leaf L is a separatrix by our definition. Q.E.D.

Let $\{S_i\}$ be a sequence of disjoint chords in $\bar{\Omega}$. We say that $\{S_i\}$ converge to a chord S, and denote this by $S_i \to S$, if both endpoints of S_i converge to the corresponding endpoints of S, respectively when $i \to \infty$. The chord S is called a *limit chord.* We note that this condition is the same as $\sigma(S_i, S) \to 0$, where for the distance function σ on arcs; see pp. 248-249 [Wh].

Lemma 1.5.6. *Let* $\{S_i\}$ *be a sequence of chords in* $K(F)$ *such that* $S_i \to S$. *Let* $S_i = \Delta(L_i)$ *for leaves* $\{L_i\}$ *in* F. *Suppose there exists a leaf* L *in* F *such that* $\Delta(L) = S$. *Then* $L_i \to L$ *in the coarse leaf topology.*

Proof. Take a pair of leaves L^1, L^2 such that L is contained in the open region U bounded by L^1 and L^2. Then S is contained in the open region $\Delta(U)$ bounded by $\Delta(L^1)$ and $\Delta(L^2)$. The condition $S_i \to S$ implies that all but finitely many S_i are contained in $\Delta(U)$. Thus all but finitely many L_i are contained in U. Q.E.D.

We note that the converse of Lemma 1.5.6 is not true. Roughly speaking, the bijection Δ^{-1} is "continuous" but Δ is not. To see this, let P be a polygonoid with the open side S_0 and closed sides $S_1, S_2, \ldots$. There exists a sequence of chords $\{C_i\}$ in $K(F)$ converging to S_0. Let $L_i = \Delta^{-1}(C_i)$ and let $L^j = \Delta^{-1}(S_j)$ be the corresponding separatrices. Then the sequence of leaves $\{L_i\}$ converges to L^j for every j. In other words, we have $L_i \to L^j$ when $i \to \infty$ for every j, but $\Delta(L^j) \notin \overline{\{\Delta(L_i)\}}$ for all j. See Fig. 1.3.

The converse of Lemma 1.5.5 is also false. For such an example, consider the limit leaf in Fig. 1.10, where we foliate the lower part of the disc as the upper one. However any such counterexamples have to be of this type, as the following theorem indicates.

Theorem 1.5.7. *A leaf* L *of* (Ω, F) *is a separatrix iff one of the following holds:*

(1) $\Delta(L)$ *is a closed side of a polygonoid of* $K(F)$.

(2) *There is an infinite sequence of distinct polygonoids* $\{P_i\}$ *in* $K(F)$ *such that there is a closed side* S_i *of* P_i *for each* i *with* $S_i \to \Delta(L)$ *(cf. Fig. 1.12(b)).*

Proof. Suppose (1) holds. Then L is a separatrix by Lemma 1.5.5. Suppose (2) holds. Then we have separatrices $L_i = \Delta^{-1}(S_i)$ for $i = 1,2,\ldots$. By Lemma 1.5.6 we have $L_i \to L$ in the coarse leaf topology. Thus L is a separatrix by 1.5.2.

Conversely, assume $\Delta(L)$ does not satisfy either (1) or (2). Since the chordal system $K(F)$ is maximal in a certain sense (Definition 1.4.5), $\Delta(L)$ is contained in a region U bounded by two chords C_1, C_2 in $K(F)$. Let $\bar{U} = C_1 \cup C_2 \cup U$. The set $\Delta^{-1}(\bar{U})$ in the leaf space is thus closed in the coarse leaf topology.

Since $\bar{U}$ contains no cyclic triples of leaves, by Lemma 5.1 and Theorem 5.1 [Mar], $\Delta^{-1}(\bar{U})$ is parallel, i.e. conjugate to a trivial product foliation. Thus the leaf L is closed in $\Delta^{-1}(\bar{U})/F$ and also closed in Ω/F. So L is not a separatrix. Q.E.D.

Recall that a chord $\Delta(L)$ satisfying condition (2) in Theorem 1.5.7 is called a limit chord.

Corollary 1.5.8. *Let* 0 *be a canonical region of* (Ω, F). *Then* $\Delta(0)$ *is one of the following five types of regions in the Kaplan diagram* $K(F)$ *(see Fig. 1.13):*

(i) *an abutment (1.4.3);*

(ii) *the interior of the union of a polygonoid with the closure of the region bounded by the open side and an arc of* $\partial\Omega$ *which is filled up by chords in* $K(F)$;

(iii) *the same as (ii) except that the arc of* $\partial\Omega$ *is replaced by* $\Delta(L)$, *for some separatrix* L;

(iv) *the interior of the union of two polygonoids with the closure of the region bounded by their two open sides which is filled up by chords in* $K(F)$;

(v) *the region between two chords* $\Delta(L_1)$ *and* $\Delta(L_2)$ *in* $K(F)$, *both* L_1 *and* L_1 *are separatrices, and the region again is filled up by chords in* $K(F)$.

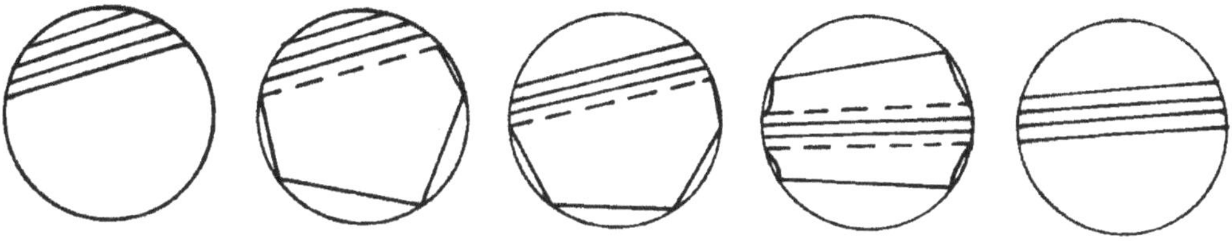

(i) (ii) (iii) (iv) (v)

Fig. 1.13

Example 1.5.9. A warning is that *not* every polygonoid is contained in $\Delta(0)$ for some canonical region 0. The polygonoid P shown in Fig. 1.13(a) is such a case. The two closed sides $\Delta(L^1)$ and $\Delta(L^2)$ of P are not limit chords, but $L_n \to L^i$, $i = 1,2$, with respect to both the coarse leaf topology and the quotient topology.

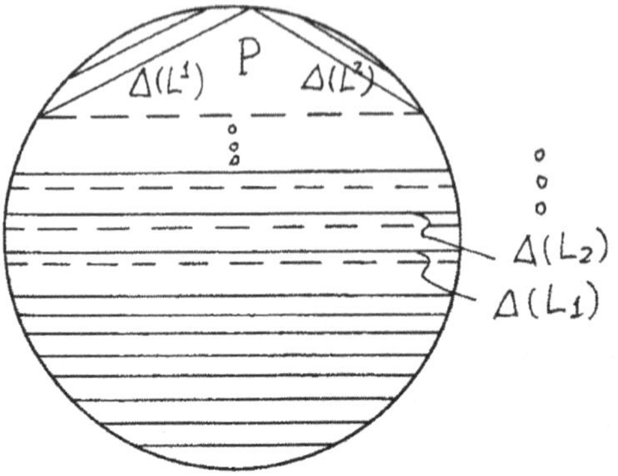

Fig. 1.13(a)

Example 1.5.10. There exist "extraordinary" foliations of the plane in which the sets of separatrices have positive planar measures under any topological conjugacies. Consider the following example (this was suggested by M. Hirsch).

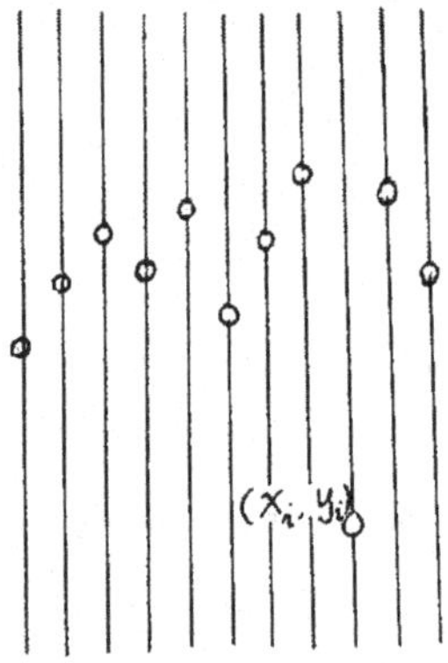

Given the trivial foliation of $\mathbb{R}^2$ by vertical constant leaves, we puncture countably infinitely many holes: $\{(x_i, y_i)\}_{i \in \mathbb{Z}}$, such that

(1) the points $\{(x_i,y_i)\}$ are discrete, (2) the horizontal coordinates $\{x_i\}_{i\in\mathbb{Z}}$ are dense in $\mathbb{R}$. Denote this foliation of punctured plane by $(\Omega_\infty, F_\infty)$.

Now let (Ω, F) be the foliated Ω as the universal covering space of $(\Omega_\infty, F_\infty)$. Then one can check that every leaf in (Ω, F) is a separatrix. To see this, we notice that (1) in $(\Omega_\infty, F_\infty)$, each of the two leaves of $(\{x_i\}\times\mathbb{R}\setminus\{(x_i,y_i)\})$ is a separatrix [Ma]. Therefore the separatrices are dense in $(\Omega_\infty, F_\infty)$ and consequently every leaf of $(\Omega_\infty, F_\infty)$ is a separatrix. (2) The preimage $\pi^{-1}(L)$ of every leaf L in $(\Omega_\infty, F_\infty)$ is a countable union of leaves, each of them being mapped bijectively onto L. The leaves in $\pi^{-1}(L)$ are separatrices in (Ω, F) iff L is a separatrix.

Proposition 1.5.11. *Suppose the set* S *of separatrices of a foliation* (Ω, F) *is Hausdorff as a subset in* Ω/F *with respect to the quotient topology. Then every polygonoid* P *in the Kaplan diagram* $K(F)$ *is contained in some* $\Delta(0)$ *for a canonical region* 0.

Proof. If a polygonoid P is not contained in a $\Delta(0)$ for a canonical region 0, then there is no region adjacent to the open side S of P filled up by parallel chords, and there are closed sides S_i of polygonoids P^i converging to S_0. By Theorem 1.5.7, the leaves $L_i = \Delta^{-1}(S_i)$ are separatrices. It is easy to see that the sequence $\{L_i\}$ converges to $L^j = \Delta^{-1}(S^j)$ in the quotient topology for any closed side S^j of P. In fact the areas of regions U_i in Ω bounded by $\{L^j\}$ and L_i converge to zero as $i\to\infty$. With respect to either the quotient topology or the coarse leaf topology, the set S is not Hausdorff.

Q.E.D.

We shall see that the converse of Proposition 1.5.11 is also true, although it is less obvious. Motivated by the structures of C*-algebras, we introduce:

Definition 1.5.12. A foliation (Ω,F) is called *ordinary* if it is topologically conjugate to a foliation in which the set S of leaves has planar measure zero. A foliation is said to be of *composition length* n, for some $n \in \{1,\dots,\infty\}$ if (1) the set S if separatrices can be expressed as the union of the nest of subsets $\emptyset = S_0 \subset S_1 \subset S_2 \subset \cdots \subset S_n = S$, where the S_i are open subsets of S and $S_i \setminus S_{i-1}$ is Hausdorff with respect to the restricted quotient topology, but S cannot be expressed as a union of such a nest of shorter length (cf. §4.4).

When S is Hausdorff, we call (Ω,F) a *foliation with* T_2*-separatrices.*

Clearly the foliation with T_2-separatrices are exactly the foliations of composition length 1 which include in particular all foliations without limit separatrices. We also note that the planar measure of a set of leaves is not quasi-invariant under topological conjugacy. So it is legal for an ordinary foliation to have some separatrices with positive planar measures. The point is that the "number" of separatrices of an ordinary foliation is negligible (although it may be uncountable) compared with that of the leaves in canonical regions, so that the separatrices in an ordinary foliation are "separating" canonical regions, as they were designed.

1.5.13 In the remainder of this paper, we shall be mostly concerned with foliations with T_2-separatrices, which will be referred to simply as "foliations" unless otherwise stated.

§2. Various trees and graphs

"Graphs" and "trees" are among the most easily visualized concepts in mathematics. However due to the wide applications of this subject in a vast number of fields, the same term may have slightly different meaning in different situations. For preciseness, it is necessary to fix some terminology.

The "trees" studied in combinatorics such as graph theory usually have geometric realizations, being connected and simply connected simplicial 1-complexes. There is a slightly more general type of "tree" as defined in J. P. Serré's beautiful book [S]. The geometric realizations of these "trees" are connected and simply connected (infinite) CW 1-complexes. These will be called "ordinary trees".

Recently still more general types of "trees" have been studied by topologists, geometers and group theorists. J. Morgan and P. Shalen studied "Λ-trees" for an arbitrary ordered abelian group Λ [M-S1]. When $\Lambda = \mathbb{R}$, these are "$\mathbb{R}$-trees" or "generalized trees" studied a little earlier by Alperin-Moss and Chiswell ([A-M], [Chi]). When $\Lambda = \mathbb{Z}$ we obtain the ordinary trees.

The distinguished trees which we introduce are a class of $\mathbb{R}$-trees. Quite often they are reduced to ordinary trees, or even finite trees. Totally unaware of the above works of the topologists and geometers, we came across the concept of $\mathbb{R}$-trees from an entirely different angle. There is a certain connection between the Morgan-Shalen construction of $\mathbb{R}$-trees for measured geodesic laminations on surfaces and the construction of our distinguished trees (see the introduction to §3).

Unless otherwise stated, all the metric spaces in this section are separable.

2.1. Ordinary graphs and trees

An excellent reference for this section is J-P. Serrê's book [S].

Definition 2.1.1. An *ordinary graph*, or for simplicity *graph*, Γ consists of a set $V = \text{ver}\,\Gamma$, a set $E = \text{edge}\,\Gamma$ and two maps

$$E \longrightarrow V \times V, \quad y \longmapsto (O(y), t(y))$$

and

$$E \longrightarrow E, \quad y \longmapsto \bar{y}$$

which satisfy the following conditions: for each $y \in E$ we have $\bar{\bar{y}} = y$, $\bar{y} \neq y$ and $O(y) = t(\bar{y})$. An element $P \in V$ is called a *vertex* of Γ; an element $y \in E$ is called an *edge*, and $\bar{y}$ is called the *inverse edge*. The vertices $o(y)$ and $t(y)$ are called the *origin* and the *terminus* of y, respectively. Together they are the *extremities* of y. Two vertices are *adjacent* if they are the extremities of the same edge. The definitions of *morphisms* and *isomorphisms* for graphs are obvious.

Definition 2.1.2. Let Γ be a graph and let $V = \text{vert}\,\Gamma$, $E = \text{edge}\,\Gamma$. Form the topological space T which is the disjoint union of V and $E \times [0,1]$. The *geometric realization* or just *realization* of the graph Γ is the quotient space $\text{real}(\Gamma) = T/R$, where the equivalent relation R is given by $(y,t) \equiv (\bar{y},1-t)$, $(y,0) \equiv o(y)$, and $(y,1) \equiv t(y)$ for $y \in E$ and $t \in [0,1]$.

We note that $\text{real}(\Gamma)$ is a CW-complex of dimension ≤ 1. Realization is a functor which commutes with direct limits.

Definition 2.1.3. A graph Γ is *connected* if $\text{real}(\Gamma)$ is a connected topological space. A nonempty graph Γ is called a *forest* if the fundamental group $\text{real}(\Gamma)$ is trivial. A connected forest is an *ordinary tree* or *tree*.

Example 2.1.4. Let $\text{ver}(\Gamma) = \{-\frac{1}{n};\ n = 1,2,3,\ldots\} \cup \{0,1\}$, $\text{edge}(\Gamma) = \{[-\frac{1}{n},-\frac{1}{n+1}],[-\frac{1}{n+1},-\frac{1}{n}];\ n = 1,2,\ldots\} \cup \{[0,1],[1,0]\}$. Then Γ is not connected. Notice that a graph Γ is connected iff $\text{real}(\Gamma)$ is connected.

Definition 2.1.5. Let p be a vertex of a graph Γ. Let V_p be the set of edges y such that $p = t(y)$. Then the cardinality n of V_p is called the *incidence number of* p, denoted by $\text{Inc}(p)$.

There is an elementary but useful fact for finite graphs.

Proposition 2.1.6. *Let* Γ *be a connected graph with a finite number of vertices. Let* $s = \text{card}(\text{ver}\ \Gamma)$ *and* $a = \frac{1}{2}\,\text{card}(\text{edge}\ \Gamma)$. *Then* $a \geq s-1$ *and equality holds iff* Γ *is a tree.*

Proof. It is easy (see p.21 [S]). Q.E.D.

2.2. **Λ-trees**

In this section we introduce the "Λ-trees" defined by Morgan and Shalen, and we shall also define certain refined structures for our needs.

2.2.1 Let Λ be an *ordered abelian group*, i.e., an abelian group together with a total ordering $<$ subject to the rule $a < b$ and $c \leq d$ implies $a+c < b+d$. The most important examples of ordered abelian groups are the real $\mathbb{R}$ and its subgroups.

2.2.2. If we replace $\mathbb{R}$ by an ordered abelian group Λ in the axioms of a real metric space, we obtain the definition of a *Λ-metric space*. A *closed interval* in Λ is a subset $[\lambda_1, \lambda_2] = \{\lambda \in \Lambda \mid \lambda_1 \leq \lambda \leq \lambda_2\}$. The *length* of the interval $[\lambda_1, \lambda_2]$ is defined to be $\lambda_2 - \lambda_1$. An *interval* in Λ is a subset I of Λ such that $a, b \in I$ and $a \leq b$ implies $[a,b] \subset I$. An *endpoint* of an interval I is the greatest or least element of I if it exists. The *length* of an interval I is the upper bound of the lengths of closed subintervals of I.

2.2.3. Let X be a Λ-metric space. A (closed) *segment* in X is a subset of X isometric to a (closed) interval in Λ. A segment is *nondegenerated* if it contains at least two distinct points. A specific isometry of a (closed) interval in Λ onto a segment is a *parametrized (closed) segment*. If $\mu: I \longrightarrow X$ is a parametrized segment, then the *length* of the segment $\mu(I)$ is the length of I. A (closed) arc in X is a subset of X which is homeomorphic to a (compact) interval in Λ. We denote by $\bar{\mu}$ the segment given by μ.

Definition 2.2.4 (II.1.2 [M-S]). A *Λ-tree* is a Λ-metric space (T,d) satisfying the following conditions:

(i) The metric space T is uniquely connected by segments, i.e., any two points of T are the endpoints of a unique closed segment.

(ii) If two closed segments in T have a common endpoint, then their intersection is a closed segment (possibly degenerated).

(iii) If $a \leq c \leq b$ are in Λ and if $\mu: [a,b] \longrightarrow T$ is a map such that $\mu|[a,c]$ and $\mu|[c,b]$ are parametrized segments and $\mu([a,c]) \cap \mu([c,b]) = \{\mu(c)\}$, then μ is a parametrized segment.

2.2.5. Condition (iii) above can be replaced by the following condition:

(iii)$'$ Under the hypotheses of condition (iii) we have $d(\mu(a),\mu(b)) = b - a$.

For $\Lambda = \mathbb{R}$ condition (ii) is superfluous.

2.2.6. The most important types of Λ-trees are those such that Λ is either $\mathbb{R}$ or a subgroup (discrete or not) of $\mathbb{R}$. Roughly speaking, when Λ is a discrete subgroup of $\mathbb{R}$, then one obtains the class of $\mathbb{Z}$-trees, which are actually the class of ordinary trees. When Λ is a nondiscrete (so necessarily dense) subgroup of $\mathbb{R}$, then a Λ-tree T can be uniquely isometrically embedded in an $\mathbb{R}$-tree $T_{\mathbb{R}}$ which is the completion of the metric space T. More precisely we have

Theorem 2.2.7 (Theorem II.1.9 [M-S]). *Let Λ be a subgroup of $\mathbb{R}$. Then every Λ-tree T can be embedded isometrically in an $\mathbb{R}$-tree $T_{\mathbb{R}}$. Moreover, $T_{\mathbb{R}}$ may be chosen so that* (a) *$T_{\mathbb{R}}$ is a complete metric space, and* (b) *the union of all closed segments in $T_{\mathbb{R}}$ with endpoints in T is a dense subset of $T_{\mathbb{R}}$. The $\mathbb{R}$-tree $T_{\mathbb{R}}$ containing T and satisfying* (a) *and* (b) *is unique in the sense that if $T_{\mathbb{R}}$ and $T'_{\mathbb{R}}$ are two such $\mathbb{R}$-trees, then the identity map of T extends uniquely to an isometry between $T_{\mathbb{R}}$ and $T'_{\mathbb{R}}$. If Λ is dense in $\mathbb{R}$ then T is dense in $T_{\mathbb{R}}$. If Λ is discrete and generated by $\ell > 0$, then $T_{\mathbb{R}}$ may be identified homeomorphically with a (geometric) 1-connected* *simplifical* 1*-complex whose* 0*-skeleton is* T *and each edge of* $T_{\mathbb{R}}$ *is then a closed segment of length* ℓ.

Thus $\mathbb{R}$-trees are the most important kind. They were studied earlier than [M-S] by R. Alperin and K. Moss and I. Chiswell under the name of "generalized trees". There is a topological characterization of $\mathbb{R}$-trees, again due to Morgan and Shalen.

Proposition 2.2.8 (Proposition II.1.13 [M-S]). *A metric space* T *is an* $\mathbb{R}$*-tree iff for any pair of points* x, y *in* T *there is a unique closed segment* I *connecting them.*

The above proposition can be used as an alternative definition, which is more convenient. For any two points x, x' in an $\mathbb{R}$-tree (T,d), we often write $I_{x,x'}$, the unique closed segment connecting them.

Example 2.2.9. Let

$$X = \{(x,y) \in [0,1] \times [0,1] \mid \text{if } x \text{ is rational than } y = 0\}$$

with the lengths of arcs given by the Euclidean metric d inherited from $\mathbb{R}^2$. Then X is an $\mathbb{R}$-tree but not an ordinary tree. We note that if "vertices" still make sense, they are dense in the interval.

2.3. Regular $\mathbb{R}$-trees

Unlike ordinary trees, an $\mathbb{R}$-tree in general does not have vertices and edges specified *a priori* in Morgan-Shalen's definition. Recall Example 2.2.9. However for our purpose, it is necessary to attach such extrastructures to $\mathbb{R}$-trees.

Lemma 2.3.1. *Let* (T,d) *be an* $\mathbb{R}$*-tree. Fix any* $x \in T$*. Suppose* I_1*,* I_2 *and* I_3 *are closed segments in* T *with* x *as a common endpoint such that both* $I_1 \cap I_2$ *and* $I_2 \cap I_3$ *are nondegenerate (closed segments). Then* $I_1 \cap I_3$ *is also a nondegenerate closed segment.*

Proof. Let $\mu_i\colon [0,\ell_i] \to T$ be a parametrized closed segment such that $\mu_i([0,\ell_i]) = I_i$, $\mu_i(0) = x$, $i = 1,2,3$. Since $I_1 \cap I_2$ is nondegenerate, there is a parametrized closed segment $\mu\colon [0,\ell] \to T$ such that $\mu([0,\ell]) = I_1 \cap I_2$ and $\mu(0) = x$. For each $\lambda \in [0,\ell]$, $\mu(\lambda)$ is the unique point in $I_1 \cap I_2$ such that the distance from $\mu(\lambda)$ to x is exactly λ. This is also true for both μ_1 and μ_2. Thus $\mu(\lambda) = \mu_1(\lambda) = \mu_2(\lambda)$ for $\lambda \in [0,\ell]$. We apply the same argument to I_2 and I_3 and get some $\ell' > 0$ such that $\mu_2(\lambda) = \mu_3(\lambda)$ for $\lambda \in [0,\ell']$. Hence $\ell_0 = \min(\ell,\ell') > 0$ and for $0 \le \lambda \le \ell_0$, we have $\mu_1(\lambda) = \mu_2(\lambda) = \mu_3(\lambda)$ and $\mu_i([0,\ell_0]) \subset I_1 \cap I_2 \cap I_3 \subset I_1 \cap I_3$ for $i = 1,2,3$. Thus $I_1 \cap I_3$ is nondegenerate. Q.E.D.

2.3.2. Let T be a nondegenerate $\mathbb{R}$-tree, i.e., T has at least two distinct points. If T is not complete, take its completion, so we shall always assume it is complete. For each $x \in T$ we let

$$F_x = \{\text{parametrized segment } \mu\colon [0,\ell] \to T \mid \mu(0) = x,\ \ell > 0\}.$$

Define an equivalence relation $\sim$ on F_x by stipulating that $\mu_1 \sim \mu_2$ iff the corresponding closed segments have nondegenerate intersection, i.e., the intersection is a nondegenerate segment. By Lemma 2.3.1 it is easy to see that $\sim$ is indeed an equivalence relation.

Denote $E_x = F_x/\sim$. An element of E_x is called a *geodesic direction* at x. The cardinal of E_x is called the *incidence number* of x and is denoted by $\mathrm{Inc}(x)$. Since X is connected, $\mathrm{Inc}(x) \geq 1$ for any x in T unless T is degenerated to a point. Let $B_T = \{x \in T \mid \mathrm{Inc}(x) \geq 3\}$ and call a point in B_T a *branch point*. We let $P_T = \{x \in T \mid \mathrm{Inc}(x) = 1\}$ and call a point in P_T a *pending point* or a *terminal point*.

Definition 2.3.3. A *regular* $\mathbb{R}$-*tree* is a triplet (T,d,V), where (T,d) is a nondegenerate $\mathbb{R}$-tree and V is a discrete subset of T containing $B_T \cup P_T$ (cf. 2.3.2), and with the property that for every $x \in T$ each connected component of $T \setminus \{x\}$ contains at least one point in V. The set V is called the *set of vertices*. Let E be the union of the sets E_x, $x \in V$, as defined in 2.3.2. Then the set E is called the *set of edges*.

Note that if (T,d,V) is a regular $\mathbb{R}$-tree, then V is countable since T is separable. Also $B_T \cup P_T$ is countable and discrete. Because of the importance of the set of edges, we often write (T,d,V,E) for a given regular $\mathbb{R}$-tree.

We shall let $\bar{V}$ denote the closure of V in T, and let $V' = \bar{V} \setminus V$.

Example. Let $T = [0,1)$ with the usual metric d, and let $V = \{0\}$. Then (T,d,V) is *not* a regular $\mathbb{R}$-tree because the vertices are not sufficiently "abundant".

Lemma 2.3.4. *Let* (T,d,V,E) *be a regular* $\mathbb{R}$-*tree. Then there is an injective map* $\gamma\colon E \to V \times \bar{V}$ *such that for each* $x \in V$, $y \in E_x$, $\gamma(y)$ *is an ordered pair* $(x, x_y) \in V \times \bar{V}$. *Moreover, let* $\mu \in F_x$ *be the unique closed parametrized segment* $\mu\colon [0,\ell] \to T$ *such that* $\mu(\ell) = x_y$. *Then* (i) $[\mu] = y$ *and* (ii) $\mu((0,\ell)) \cap V = \emptyset$.

Conversely, if $\mu \in F_x$ *such that* $\mu: [0,\ell] \to T$, *with* $\mu(\ell) \in \bar{V}$ *and* $\mu((0,\ell)) \cap V = \emptyset$, *then* $\mu(\ell) = x_y$ *for* $y = [\mu] \in E_x$.

Furthermore if $x_y \in V$ *for some* $\gamma(y) = (x,x_y)$, *then there is a unique* $\bar{y} \in E_{x_y}$ *such that* $\gamma(\bar{y}) = (x_y,x)$. *We shall call* $\bar{y}$ *the inverse of* y.

Proof. Let $\nu: [0,r] \to T$, $\nu \in F_x$ such that

(1) $[\nu] = y$;

(2) $\nu((0,r])$ contains at least one element $x' \in V$.

Such ν exists by the definition of a regular $\mathbb{R}$-tree.

Put $\ell = \inf\{t \in (0,r] \mid \nu(t) \in V\}$. Then $\ell > 0$ as V is discrete. Let $x_y = \nu(\ell)$ and $\mu = \nu|[0,\ell]$.

First, $\mu([0,\ell])$ is obviously the unique closed segment connecting x and x_y. Next, from our construction $[\mu] = [\gamma] = y$ and $\mu(0,\ell) \cap V = \emptyset$. It remains to show that $\gamma(y) = (x,x_y)$ is well-defined. Let $\nu' \in F_x$ and assume that $\nu': [0,\ell'] \to T$ also satisfies (1) and (2). Suppose that the above construction yields $x'_y = \nu'(\ell')$, with $x'_y \neq x_y$. Let $r_0 = \sup\{t \in [0,r] \mid \nu(t) = \nu'(t)\}$. Then $r_0 > 0$ since $[\nu] = [\nu'] = y$. On the other hand, $r_0 < \ell$ for otherwise from the way we defined ℓ' we would have $\ell' = \ell$ and $x_y = x'_y$.

Let $x_0 = \nu(r_0) = \nu'(r_0)$. Define parametrized segments:

$$\mu_1(t) = \nu(r_0-t) \quad \text{for } 0 \leq t \leq r_0$$

$$\mu_2(t) = \nu(t+r_0) \quad \text{for } 0 \leq t \leq \ell-r_0$$

$$\mu_3(t) = \nu'(t+r_0) \quad \text{for } 0 \leq t \leq \ell'-r_0 .$$

One verifies that $[\mu_1]$, $[\mu_2]$, $[\mu_3]$ are all different in E_x, so $x_0 \in B_T \subset V$. Since $x_0 = \nu(r_0)$, this again contradicts the way ℓ

was chosen.

The rest of the assertion is now clear. Q.E.D.

Theorem 2.3.5. *Let* (T,d,V,E) *be a regular* $\mathbb{R}$*-tree. Assume that the set* V *of vertices is closed. Then* (V,E) *is an ordinary tree (Definition 2.1.3) and* (T,d,V,E) *can be identified with the geometric realization of* (V,E).

Proof. We assume the notations in Lemma 2.3.4. When the set V of vertices is a closed subset of T, we have $x_y \in V$ for every $y \in E_x$ and every $x \in V$, so the map γ is an injection from the set E into $V \times V$. For $y \in E$ such that $\gamma(y) = (x, x_y)$, let $\bar{y} \in E$ so that $\gamma(\bar{y}) = (x_y, x)$. Clearly (V,E) is a graph (Definition 2.1.1).

Now for any ordered pair $(x,x') \in V \times V$, there is a unique closed parametrized segment $\mu\colon [0,\ell] \to T$ with $\mu(0) = x$, $\mu(\ell) = x'$, $0 < \ell$. Clearly the set $\mu[0,\ell] \cap V$ is finite, since its preimage is discrete and closed subset of $[0,\ell]$. Let $\mu[0,\ell] \cap V$ consist of $x_0 = x, x_1, \ldots, x_n = x'$ in its natural order. Then these vertices and edges $\{[x_i, x_{i+1}], [x_{i+1}, x_i]\}$ form a path connecting x and x'. So (V,E) is a tree. It is now clear that (T,d,V,E) can be viewed as the geometric realization of the tree (V,E), up to homeomorphism. Q.E.D.

2.3.6. Two Λ-trees (T,d) and (T',d') are *isometric* or *isomorphic* if there exists a bijection $\phi\colon T \to T'$ such that $d'(\phi(x_1),\phi(x_2)) = d(x_1,x_2)$. The mapping ϕ is then called an *isometry*. Two $\mathbb{R}$-trees (T,d) and (T',d') are *similar* if there is a homeomorphism $\sigma\colon T \to T'$. The same terms *isomorphic* or *similar* apply to regular $\mathbb{R}$-trees if in addition the corresponding isometry ϕ or homeomorphism σ induces a bijection between the sets of vertices and therefore the sets of edges.

2.4. Quotient graphs and quotient $\mathbb{R}$-trees

In §3 we shall introduce "distinguished trees" for foliations of the plane. Kaplan also introduced "generalized trees" in his definition of normal chordal systems (p.173 [K1]). (We shall call these the *Kaplan trees*.) A Kaplan tree is actually an ordinary tree. The Kaplan trees are highly noncanonical in the sense that for a given foliation (Ω,F) or a chordal system there are in general infinitely many noniso-morphic Kaplan trees corresponding to it. However it turns out that they are all certain "duals" of quotient trees of the distinguished tree which is uniquely associated to the foliation (see 3.5.5).

To prepare for the discussion in the next chapter, we need to make the concepts of quotient graphs and quotient trees precise. For ordinary graphs, this can be found in (p.22 [S]).

2.4.1. Let Γ be a connected nonempty graph. Let $S = \cup S_i$ be a subgraph of Γ which is a disjoint union of subtrees S_i. The *quotient graph* Γ/S is defined as follows.

The set of vertices of Γ/S is the quotient of $\operatorname{vert}\Gamma$ by the equivalence relation whose classes are the sets $\operatorname{vert} S_i$ and the individual elements in $\operatorname{vert}\Gamma - \operatorname{vert} S$. The set of edges of Γ/S is $\operatorname{edge}\Gamma - \operatorname{edge} S$, with the involution $y \mapsto \bar{y}$ induced from that in $\operatorname{edge}\Gamma$. The mapping

$$\operatorname{edge}(\Gamma/S) \longrightarrow \operatorname{vert}(\Gamma/S) \times \operatorname{vert}(\Gamma/S)$$

is induced by

$$\text{edge } \Gamma \to \text{vert } \Gamma \times \text{vert } \Gamma$$

by passing to quotients.

It is easy to see that real(Γ/S) is the quotient space of real(Γ) obtained by contracting each real(S_i) to a point.

2.4.2. Let (T,d) be an $\mathbb{R}$-tree. An $\mathbb{R}$-subtree (S_1,d_1) is a metric subspace of T which is an $\mathbb{R}$-tree with the induced metric $d_1 = d|S_1$.

If S_1, S_2 are $\mathbb{R}$-subtrees of (T,d) then $S_1 \cap S_2$ with the induced metric is also an $\mathbb{R}$-subtree of (T,d) as long as it is not empty. In fact, for a pair of points x, x' in $S_1 \cap S_2$, there is a unique closed segment I_i in S_i connecting them for $i = 1,2$. However, both I_1, I_2 are also unique as closed segments in T, so I_1 and I_2 coincide and are contained in $S_1 \cap S_2$.

In particular, if S_1 is a subtree of (T,d) and S_2 is an arc such that $S_1 \cap S_2 \neq \emptyset$, then $S_1 \cap S_2$ is also an arc.

2.4.3. Let (T,d) be an $\mathbb{R}$-tree. Let $\{S_i\}$ be a family of disjoint closed sub $\mathbb{R}$-trees of (T,d) such that every compact subset of T intersects at most finitely many S_i. Let $S = \cup S_i$.

We introduce a metric $\bar{d}$ on the quotient space T/S as follows. For any $\bar{x}_1$, $\bar{x}_2$ in T/S, choose x_1, x_2 in T with $x_1/S = \bar{x}_1$ and $x_2/S = \bar{x}_2$. There is a unique closed segment I connecting x_1 and x_2. By 2.4.2 the $I \cap S_i$ are all closed segments and only a finite number of them are nonempty. Now we define

$$\bar{d}(\bar{x}_1,\bar{x}_2) = |I| - \sum_i |I \cap S_i| \; . \tag{2.1}$$

It is easy to see that $\bar{d}$ is well-defined.

We claim that $\bar{d}$ defines a metric on T/S. For let $\bar{d}(\bar{x}_1,\bar{x}_2) = 0$. Then $|I| = \sum_i |I \cap S_i|$ by (2.1). This is a finite sum by the hypothesis on $\{S_i\}$, and each $I \cap S_i$ is closed. Thus $I - \cup(I \cap S_i)$ is open in I, and so $|I| = \sum_i |I \cap S_i|$ forces it to be empty. Since the $I \cap S$ are mutually disjoint, the equality $I = \cup(I \cap S_i)$ implies that $I = I \cap S_i$ for some i. Thus $\bar{x}_1 = \bar{x}_2$.

Next let $\bar{x}_1$, $\bar{x}_2$ and $\bar{x}_3$ be in T/S. We claim that

$$\bar{d}(\bar{x}_1,\bar{x}_3) \leq \bar{d}(\bar{x}_1,\bar{x}_2) + \bar{d}(\bar{x}_2,\bar{x}_3) \ . \tag{2.2}$$

Let I, I_1 and I_2 be the closed segments connecting x_1 and x_3, x_1 and x_2, x_2 and x_3, respectively. It is easy to see $I \subset I_1 \cup I_2$. For let $\mu_1, \mu_2 \in F_{x_2}$ such that $\mu_1: [0,\ell_1] \to T$, $\mu(\ell_1) = x_1$ and $\mu_2: [0,\ell_2] \to T$, $\mu(\ell_2) = x_3$. Let $\ell_0 = \sup\{t | \mu_1(t) = \mu_2(t)\}$ and $x_0 = \mu_1(\ell_0)$. Then $I_1' = \mu_1([\ell_0,\ell_1])$ and $I_2' = \mu_2([\ell_0,\ell_2])$ are the closed segments connecting x_0 to x_1 and x_0 to x_3, respectively. Moreover $I_1' \cap I_2' = \{x_0\}$. Thus by Axiom (iii), Definition 2.2.4, $I_1' \cup I_2'$ coincides with I. Therefore $I = I_1' \cup I_2' \subset I_1 \cup I_2$. Thus this is (2.2).

One easily verifies also that the metric topology given by $\bar{d}$ coincides with the quotient topology.

The metric space $(T/S,\bar{d})$ is again an $\mathbb{R}$-tree which is called the *quotient* $\mathbb{R}$-*tree of* (T,d) *obtained by contracting* S.

In order to explain the significance of the condition that every compact subset of T intersects only finitely many S_i, let us consider a simple example.

Let $T = [0,1] \subset \mathbb{R}$. Let $\{S_i\}$ be an infinite sequence of disjoint closed subintervals in T such that $\sum |S_i| = 1$. Then (2.1) tells us that $\bar{d}(\bar{x}_1, \bar{x}_2) = 0$ for every pair $\bar{x}_1, \bar{x}_2 \in T/S$, although the quotient space T/S is uncountably infinite. Thus (2.1) fails to provide a metric for T/S.

Nevertheless, one can show that in this example T/S is compact, Hausdorff, and has a countable basis. So T/S is metrizable. In fact T/S is homeomorphic to $[0,1]$, strangely enough. Consider a continuous monotone increasing function ϕ mapping the unit interval onto itself with the property that ϕ is constant on each component of the open Cantor set. For instance, we may define ϕ as follows: $\phi(0) = 0$, $\phi(1) = 1$, $\phi([\frac{1}{3}, \frac{2}{3}]) = \{\frac{1}{2}\}$, $\phi([\frac{1}{3^2}, \frac{2}{3^2}]) = \{\frac{1}{2^2}\}$, $\phi([\frac{3+1}{3^2}, \frac{3+2}{3^2}]) = \{\frac{2+1}{2^2}\}, \ldots,$ $\phi([\frac{3^r+1}{3^k}, \frac{3^r+2}{3^k}]) = \{\frac{2^r+1}{2^k}\}, \ldots$. One checks that ϕ is continuous and increasing. Thus $\mathrm{range}(\phi) = [0,1]$ and ϕ induces a homeomorphism from T/S onto $[0,1]$. Here S is the union of the closed subintervals above.

For our purposes we restrict ourselves to the structure in which the metric on T/S is given by a simple formula (2.1), instead of striving for the utmost generality.

2.4.4. Let $T = (T,d,V,E)$ be a regular $\mathbb{R}$-tree. Let $(S, d|_S)$ be an $\mathbb{R}$-subtree of (T,d). Denote $V_S = V \cap S$. If $S = (S, d|_S, V_S)$ is a regular $\mathbb{R}$-tree by itself, then it is called a *regular* $\mathbb{R}$-*subtree of* T. If this is the case, one checks that the set of edges of S is given by $E_S = \bigcup_{x \in V_S} \{y \in E_x \mid y = [\mu] \text{ for some } \mu \in F_x \text{ with range in } S\}$ and that the set of pending points P_S is contained in V. However for any $\mathbb{R}$-subtree S, $B_S \subset B_T \subset V$ is always true.

2.4.5. Let $T = (T,d,V,E)$ be a regular $\mathbb{R}$-tree. Let (S,d) be a closed $\mathbb{R}$-subtree of (T,d) such that with $V_S = V \cap S$ as the set of vertices the triplet (S,d,V_S) is itself a regular $\mathbb{R}$-tree. Then $S = (S,d,V_S)$ is called a *regular* $\mathbb{R}$-*subtree of* T. One verifies then that $E_S = \bigcup_{x \in V_S} \{y \in E_x | y = [\mu] \text{ for some } \mu \in F_x \text{ with range in } S\}$ is the set of edges.

We remark that a necessary and sufficient condition for a closed $\mathbb{R}$-subtree (S,d) of a regular $\mathbb{R}$-tree $T = (T,d,V)$ to be a regular $\mathbb{R}$-subtree, is that the pending point set P_S of S is contained in V (see 2.3.2). In fact necessity is obvious. Suppose we have $P_S \subset V$. Then $P_S \subset V_S$. Since $B_S \subset B_T \subset V$ is always true, we have $P_S \cup B_S \subset V_S$, which is always a discrete subset of (S,d). This verifies the sufficiency.

Assume that $S = \bigcup_i (S_i,d,V_i,E_i)$ is a disjoint union of regular $\mathbb{R}$-subtrees of T satisfying the condition at the beginning of 2.4.3. Let $(T/S,\tilde{d})$ be the quotient $\mathbb{R}$-tree defined in 2.4.3. We can associate a set $\tilde{V}$ of vertices to it as follows: an element in the set $\tilde{V}$ is either a connected component of S or an element in V-S. We can define the set $\tilde{E} = \bigcup_{\tilde{x} \in \tilde{V}} \tilde{E}_{\tilde{x}}$ as usual. However the quadruplet $(T/S,\tilde{d},\tilde{V},\tilde{E})$ is in general *not* a regular tree, because V may fail to be discrete.

Example. Study the "skeleton of a dinosaur tail" as shown in Fig. 2.1(a), which is a regular $\mathbb{R}$-tree. After crushing its spine, we get (b), which is not regular.

In general we can always get a regular $\mathbb{R}$-tree $(T/S,\tilde{d},\tilde{V}_0)$ by deleting some vertices from $\tilde{V}$.

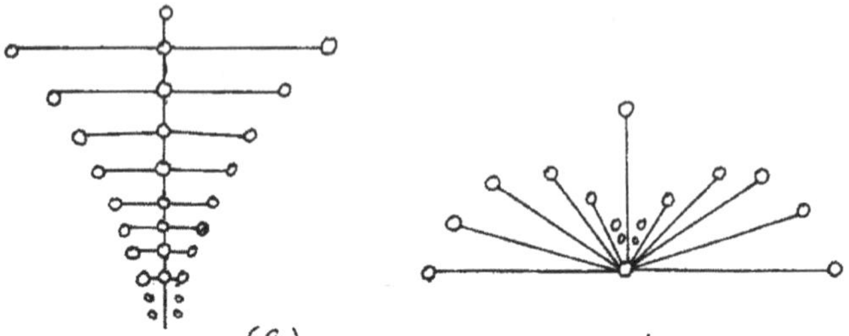

(a) (b)

Fig. 2.1

2.5. ℝ-graphs

Instead of ℝ-trees, the C*-algebras of Kaplan-Markus flows on multiconnected regions are classified by certain kinds of topological combinatorial configurations which may have nonvanishing fundamental groups. Therefore we need to generalize further the notions of ℝ-trees and regular ℝ-trees to "ℝ-graphs" and "regular ℝ-graphs".

The reader is referred to [Nagata] for definitions of dimensions on metric spaces.

Definition 2.5.1. An ℝ-*graph* is a separable metric space (X,d) satisfying the following conditions:

(i) $\dim X = 1$

(ii) Any pair of points x, y in X is connected by a closed segment (2.2.3).

Example 2.5.2. An ℝ-tree is of course an ℝ-graph. In fact an ℝ-graph X is an ℝ-tree iff axiom (ii) is strengthened to say that any pair of points x, y in X is connected by a unique closed segment. Given a finite family of ℝ-trees $\{T_i\}$ such that $\cup T_i$ is path-connected and

$T_i \cap T_j \subseteq P_{T_i} \cap P_{T_j}$, then with the metric $d(x_1,x_2) = \inf\{\sum_i |\ell \cap T_i| \mid$ are ℓ connecting x_1 and x_2 in $\cup T_i\}$, $\cup T_i$ is an $\mathbb{R}$-graph. However if $\{T_i\}$ is an *infinite* family, the conclusion may be false. Consider the example illustrated by Fig. 2.2. The metric d_i on each T_i is just given by

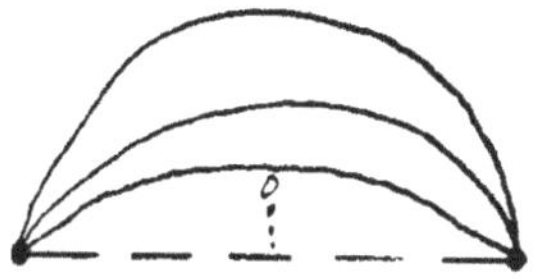

Fig. 2.2

the arc length. There is no closed *segment* connecting x_1 and x_2 in $\bigcup_i T_i$. Therefore $\bigcup_i T_i$ with the metric d defined above is *not* an $\mathbb{R}$-graph.

The proof of Lemma 2.3.1 is still valid for $\mathbb{R}$-graphs. Therefore for each $x \in X$, the sets F_x and E_x still make sense (see 2.3.2), and we can again define the set of branch points B_x and the set of pending points P_x as before.

Definition 2.5.3. A regular $\mathbb{R}$-graph is a quadruplet (X,d,V,E), where (X,d) is an $\mathbb{R}$-graph, V is a discrete subset of X containing $B_X \cup P_X$ and E is the union of sets E_x, $x \in V$. The set V is called the *set of vertices of* X and E is called the *set of edges of* X.

Example 2.5.4. Let X be the "Hawaiian earring", as shown in Fig. 2.3.

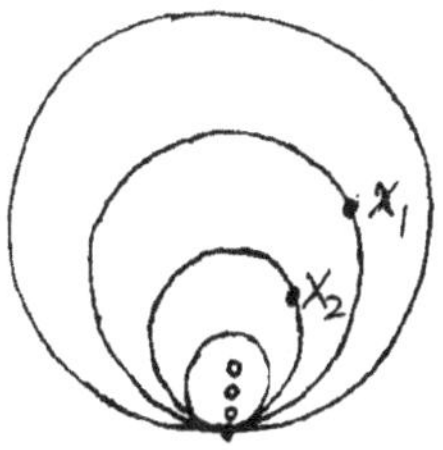

Fig. 2.3

The nth circle has radius $\frac{1}{n}$. For any two points x_1, x_2, define $d(x_1,x_2)$ to be the sum of the shorter arc length from x_i to 0, $i = 1,2$. Let $V = \{0\}$. Then (X,d,V) defines a regular $\mathbb{R}$-graph, which is not an $\mathbb{R}$-tree. Let $(\tilde{X},\tilde{d})$ be the universal covering space of (X,d) and let $\tilde{V} = \pi^{-1}(V)$, where π is the covering map. Then $(\tilde{X},\tilde{d},\tilde{V})$ defines a regular $\mathbb{R}$-tree.

§ 3 Distinguished trees

In this section we give a detailed characterization of distinguished trees which are a complete combinatorial-topological invariant of the C*-algebras of foliations with T_2-separatrices.

In §3.1 and §3.2 we construct a metric space $T(F)$ of dimension one, which is a subset of Ω but with a different metric, such that equipped with this new metric, $T(F)$ is an ℝ-tree. Endowed with some extra structure, $T(F)$ is called the tree associated to (Ω,T) , and is shown in particular to be a regular ℝ-tree (Def. 2.2.4).

In §3.2-§3.5 we give an axiomatic definition of distinguished trees. A tree associated to a foliation (Ω,F) with T_2-separatrices is shown to be a distinguished tree (Thm. 3.3.4). It is then shown that <u>every</u> distinguished tree is of the form $T(F)$, i.e., a tree associated to a foliation (Thm. 3.3.5 and Thm. 3.5.11) with T_2-separatrices.

To motivate our definition, let us consider first a very simple situation, i.e., a foliation (Ω,F) which has no limit separatrices. (The investigation in [Ma] is focused on this important special case.) We construct the dual graph of the collection of cononical regions and separatrices of (Ω,F) as follows: a vertex is a cononical region, two vertices are jointed by an (unoriented) edge if the two corresponding canonical regions share a common separatrix in their boundary. By the Jordan curve theorem, one can show easily that this graph is a tree. When limit separatrices appear, this construction fails to produce a tree. One may predict that a combinatorial invariant does not contain sufficient topological information. However, it is helpful to keep this picture in mind. In the general situation one needs to introduce topology to these tree-like structures, and we naturally obtain ℝ-trees. Although only the topology on these ℝ-trees matters, we shall find that it is convenient to have it given by a metric.

It is important to point out here that the Euclidean metric on the disc Ω is not the natural metric to consider when we view the disc as the universal covering space of a closed two surface of genus $g \leq 0$. The natural metric is, of course, the hyperbolic metric, with respect to which the action of deck transformation of the surface group is isometric. However, as we remarked earlier, not the particular metric, but the topology on the $\mathbb{R}$-trees (in particular on the sets of vertices) is invariant for the topological conjugacy class of foliations, so for the class of C*-algebras. Therefore for simplicity we may still use the Euclidean metric in defining arc length in §3.1.

For a pair of points x_1, x_2 in $\bar{\Omega}$, we denote $\overline{x_1x_2}$ the closed geodesic in $\bar{\Omega}$ connecting x_1 and x_2 with respect to the Euclidean metric. (It is the closed (straight) segment with ends x_1 and x_2, but this term confuses a little with that of "closed segments" we used in defining $\mathbb{R}$-trees in §2.)

3.1. The "nerve" of a canonical region

3.1.1. Let P be a polygonoid (Def. 1.4.2.). We shall always assume that its sides $S_1, S_2, \cdots$ have been so enumerated that their lengths are in decreasing order. Let the middle points of S_i be p_i. The middle point of the segment $\overline{p_1p_2}$ is denoted x_p and called the *center of the polygonoid P*. The union of all $\overline{x_pp_i}$, $i = 1,2,\cdots$ is called the *nerve of P* and denoted N(P). Refer to Fig. 3.1. Of course both x_p and N(P) depend on the enumeration of S_1 and S_2, which may fail to be unique for certain polygonoids P. However this ambiguity is not serious because most polygonoids in a Kaplan diagram are "long" and "thin". Later we can easily see that up to similarity, the "regular $\mathbb{R}$-tree" $T(F)$ obtained by gluing all the N(P) together is independent

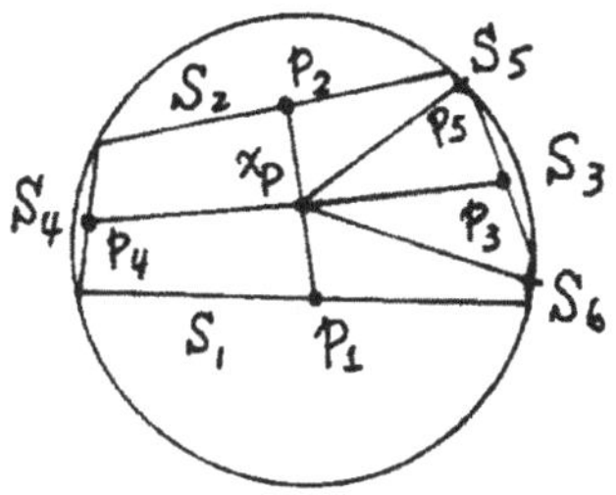

Fig. 3.1

of a particular choice of $N(P)$ and x_p for these exceptional polygonoids P in the Kaplan diagram $K(F)$. Therefore we may assume that the polygonoid P is of generic type and S_1, S_2 have been fixed.

3.1.2. Let Δ be the Kaplan map (1.2.4). Let O be a canonical region of (Ω, F). There are five possible types of $\Delta(O)$ in the Kaplan diagram $K(F)$, which are given in Corollary 1.5.8. We define a one-dimensional topological subspace of $\Delta(O)$ for each O, called the *nerve of O* (or *of* $\Delta(O)$) and denoted by $N(O)$. The construction of $N(O)$ is described separately as follows (see Fig. 3.2). For notations see Corollary 1.5.8 and Fig. 1.13.

Case 1) Suppose $Q = \Delta(O)$ is of type (i). Recall that Q is bounded by a closed side, say $\overline{x_1x_2}$ of a polygonoid, and a circumference arc $\widehat{x_1x_2} \subset \partial\Omega$. Let p_1 and p_2 be the middle point of $\overline{x_1x_2}$ and $\widehat{x_1x_2}$ respectively. The middle point of $\overline{p_1p_2}$ is denoted by x_Q and is called the *center of Q* (or *of O*). The nerve $N(O)$ of O is just the closed segment $\overline{x_Qp_1}$ in the metric space Ω. See Fig. 3.2.

Case 2) Suppose $\Delta(O)$ is of type (ii). Let p be the polygonoid with the open side S_k. Then the nerve $N(O)$ of $\Delta(O)$ is $(N(P) \setminus \overline{p_kx_p}) \cup \{x_p\}$ (3.1.1). The center x_o of (O) (of O) is just x_p.

Case 3) Suppose $\Delta(0)$ is of type (iii). Again let P be the polygonoid with open side S_k . The nerve N(0) is the union of N(P) with $\overline{p_k x_1}$, where x_L is the middle point of the chord $\Delta(L)$. The center x_o of $\Delta(0)$ is again x_p .

Case 4) Suppose $\Delta(0)$ is of type (iv). Let P and P' be the two polygonoids with sides $S_1, S_2, \cdots$ and $S_1', S_2', \ldots$ such that the lengths are decreasing. Let their open sides be S_k and S_1' respectively. We obtain the centers x_p, x_p' and nerves N(P) and N(P') as in 3.1.1. The nerve N(0) of 0 is the union of N(P), N(P') and $\overline{p_k p_1'}$, where p_k and p_1' are the middle points of S_k and S_1' respectively. The center of $\Delta(0)$ is the union of the two points x_p and x_p' .

Case 5) Suppose $\Delta(0)$ is of type (v) . Let the middle points of $\Delta(L_1)$ and $\Delta(L_2)$ be p_1 and p_2 respectively. Then the center of Q (or of 9) x_Q is the middle point of $\overline{p_1 p_2}$. The nerve N(0) of 0 is just $\overline{p_1 p_2}$.

The above five cases are illustrated in Fig. 3.2.

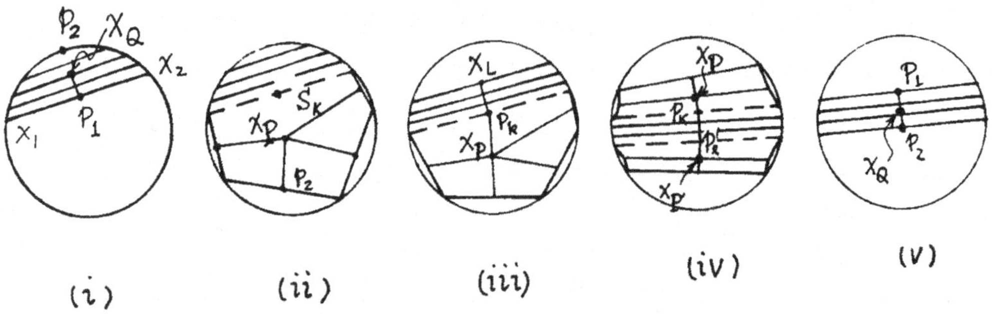

Fig. 3.2

3.1.3 A metric d is defined on each N(0) as follows. For any pair of points x_1, x_2 in N(0) there is a unique closed arc I connecting them. The arc I is the union of a few segments. Each of them has the metric induced from $\overline{\Omega}$. Let $d(x_1, x_2)$ be the sum of these segments.

Obviously the topology on N(0) defined by d is the same as that on N(0) inherited from $\overline{\Omega}$.

Lemma 3.1.4. For each canonical region O , the nerve $(N(O),d)$ is an $\mathbb{R}$-tree. The branch points of $(N(O),d)$ are among the centers of the polygonoids involved (if any).

Proof It is obvious for the cases of type (i) or (v) (there are no branch points) and of type (ii) or (iii) (there is at most one branch point). For type (vi), we note that $N(P)$ and $N(P')$ are disjoint and the segment $\overline{p_k p_1'}$ intersects with $N(P)$ and $N(P')$ only at p_k and p_1' respectively.

Q.E.D.

3.2 The associated tree of a foliation

3.2.1. Let (Ω,F) be a foliation with Kaplan diagram $K(F)$. We now define a metric space $T(F)$ associated to (Ω,F) (or to $K(F)$, equivalently). As a topological subspace of Ω , $T(F)$ is the union of all $N(O)$, for all canonical region O of (Ω,F) , together with the middle points of all $\Delta(L)$, for all limit separatrices L . (See the remarks following the proof of Lem. 1.5.5, and Thm. 1.5.7.)

Let X be any topological space. By a path in X, we mean an injective continuous map f from an interval $[a,b]$ into X , where $-\infty < a \leq b < \infty$. We say that a pair of points x_0 and x_1 are connected by a path if there is a path f defined on $[a,b]$ such that $f(a)=x_0$ and $f(b)=x_1$.

Theorem 3.2.2. The topological space $T(F)$ is path-connected, i.e., any pair of points x_0,x_1 in $T(F)$ is connected by a path. Furthermore the image of the path is uniquely determined by x_0 and x_1 .

Proof By Lem. 3.1.4, each $(N(O),d)$ is an $\mathbb{R}$-tree. In particular it is path-connected and simply connected. First we show the uniqueness of the image of a path connecting two points x_0 and x_1 , for any given pair of points x_0 and x_1 in $T(F)$. Let $x \in T(F)$ be such that x is not one of the "ends" of $T(F)$, i.e., x is not the center of any $\Delta(0)$ of type (i) (see Cor. 1.5.7 and 3.1.2). Clearly it suffices to show that $T(F)\setminus x$ is always disconnected for such x.

Note that $\Delta(L)\cap T(F)$ consists of precisely one point, i.e., the middle point of $\Delta(L)$ for any separatrix L. Since $\Delta(L)$ separates Ω into two open regions, $T(F)\setminus\{x\}$ is disconnected for any $x\in\Delta(L)\cap T(F)$. Now let x be an interior point in $N(0)$ for some canonical region 0. Let the separatrices bounding 0 be $L_1, L_2, \cdots$. Each $\Delta(L_i)$ divides $\bar{\Omega}$ into three parts: $\delta(L_i)$, $\delta'(L_i)$ and $\Delta(L_i)$. We may choose δ so that $\delta(L_i)\cap\delta(L_j)=\emptyset$, for $i\neq j$ and the intersection $\cap\delta'(L_i)$ is the interior of $\Delta(0)$. The middle point p_i of $\Delta(L_i)$ divides $T(F)$ into three: $T_i=\delta(L_i)\cap T(F)$, $T_i'= '(L_i)\cap T(F)$ and $\{p_i\}$. Correspondingly we have $T_i\cap T_j=\emptyset$ and T_i' is the interior of $N(0)$. New $N(0)\setminus x$ is disconnected, as x is an interior point. It is easy to see then that $T(F)\setminus x$ is also disconnected.

Finally we show that $T(F)$ is path-connected.

Let 0 and $0'$ be two adjacent canonical regions, i.e., the union $0\cup L\cup 0'$ is a connected region for some separatrix L. By Cor. 1.5.8 and from the construction, $N(0)\ N(0')$ always consists of a single point, i.e., the middle point of $\Delta(L)$. Thus $N(0)\cup N(0)$ is path-connected.

Let $\{L^\lambda\}_{\lambda\in\Lambda}$ be the set of limit separatrices of (Ω,F), $\{U^j\}$ the set of connected components of $\Omega\setminus\bigcup_n\Delta(L^n)$, and $T^j=U^j\cap T(F)$. For any two points $x_0,x_1\in T^j$, the straight line I in Ω connecting x_0,x_1 intersects only finitely many $\Delta(0)$, for canonical regions 0. Enumerate them by $0_1,\ldots,0_m$ according to the order of $\Delta(0)\cap I$ on I from x_0 to X_1. Then repeating the discussion above, $\bigcup_i N(0_i)$ is path-connected and contains x_0 and x_1. So T^j is path-connected.

Now let L_z be a limit separatrix, $z\in\Delta(L_z)\cap T(F)$. First we assume that L_z bounds no canonical regions, i.e., there is a sequence $\{L^k\}$ of separatrices converging to L_z from both sides. Let I be a closed straight segment in Ω orthogonal to $\Delta(L)$ and containing z in its interior. For any sequence $\{P_i\}$ of polygonoids converging to L_z, we have $\mathrm{diam}(P_i)\longrightarrow \mathrm{length}(L_z)\neq 0$, so all but finitely many P_i are acute by Thm. 1.4.4. We may then assume that I is short enough so that (1) each $\Delta(0)$ is one of the types (iii), (iv) or (v) if $\Delta(0)\cap I\neq\emptyset$, (2) for each polygonoid $P\subset\Delta(0)$ of type (iv) or (v) with $\Delta(0)\cap I\neq\emptyset$. The open side is $S_1(P)$ or $S_2(P)$, and the $S_k(P)$ do not intersect I for $k\geqslant 3$ (3.1.1). We also assume that both ends x_0, x_1 of I are on chords $\Delta(L)$'s for separatrices L's. Identifying I with an interval $[a,b]$ on the real line, we shall define a function $f: [a,b]\to\Omega$. (See Fig. 3.3, where we lowered the segment $I=[a,b]$ in order to illustrate $f(I)$. Actually both I and $f(I)$ contain z.)

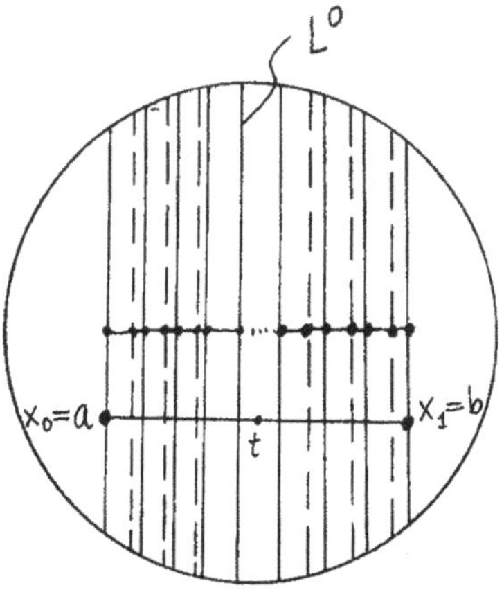

Fig. 3.3

We set $E=t\in[a,b]\mid t\in I\cap C_t\neq\emptyset$, where the chord C_t is either (1) $S_1(P)$ or $S_2(P)$ for some polygonoid P or (2) $C_t=\Delta(L)$ for some limit separatrix L. It follows from 1.5.2 that E is a closed subset of I. Note that E may be uncountable but it is always of Lebesque measure zero for all ordinary foliations.

We define $f(t)$ to be the middle point of C_t, for all $t\in E$. Note that a chord C_t may not be of form $\Delta(L)$ for some leaf L, if C_t is the open side of a polygonoid. If

$t \notin T$, then there are $t_0, t_1 \in E$ such that $t_0 < t < t_1$ and $(t_0,t_1) \cap E = \emptyset$. We define $f(t)$ by linear interpolation between $f(t_0)$ and $f(t_1)$. Thus f has been defined on entire I.

Suppose $\{t_i\} \subset E$ such that $t_i \to t_0$. Then $C_{t_i} \to C_{t_0}$ (see the remark preceding Lemma 1.5.6) and thus the middle points $f(t_i) \to f(t_0)$. It follows easily that f is continuous. f is also one-to-one since $C_t \cap C_{t'} = \emptyset$ if $t \neq t'$. So f is a path in Ω.

We claim that $f(I) \subset T(F)$. By the construction, we have $f(E) \in T(F)$. Given any $t \in I \backslash E$, we fix $t_0, t_1 \in E$ with $t_0 < t < t_1$ and $(t_0,t_1) \cap E = \emptyset$ as before. Then because of the maximum property of Kaplan diagrams ((1), 1.4.5), there are the following two possibilities:

(1) $t \in P$ for some polygonoid and so t_0, t_1 are the middle points of $S_1(P)$ and $S_2(P)$ so $f(t) \in N(P) \subset T(F)$.

(2) $t \in C$ for some chord $C = \Delta(L)$. Since $t \notin E$, the leaf L_t is not a separatrix. There are three cases:

(i) The two chords C_{t_0} and C_{t_1} bound a region of type (V) (Fig. 3.2) so $f(t) \in T(F)$.

(ii) One of C_{t_1} and C_{t_1} is the open side of a polygonoid P, as shown in Fig. 3.2(iii), where the two points x_L and P_k are just the two points $f(t_0)$ and $f(t_1)$. So $f(t) \in T(F)$.

(iii) Both C_{t_0} and C_{t_1} are open sides, as shown in Fig. 3.2(iv), where P_k and P'_ℓ are the two points $f(t_0)$ and $f(t_1)$. Again $f(t) \in T(F)$.

We have shown that for any $z \in T(F) \cap \Delta(L_z)$, where L_z is a limit separatrix bounding no canonical region, there is an arc

$T_z = f(I)$ in $T(F)$ containing z in its interior.

In the case that L_z is contained in the boundary of a canonical region, a similar argument still shows that there is an arc T_z in $T(F)$, containing $z = \Delta(L_z) \cap T(F)$ in its interior.

Let x_0, x_1 be any two points in $T(F)$. We connect them by a segment $\tau: [0,1] \to \bar{\Omega}$, $f(0) = x_0$, $f(1) = x_1$. For each $t \in [0,1]$, $f(t)$ is either in some U^j or some $\Delta(L^\lambda)$. Put $I_0 = \{t \in [0,1] \mid f(t)$ is in U^j or $f(t)$ is in $\Delta(L^\lambda)$ such that $U^j \cap T(F)$ or $\Delta(L^\lambda) \cap T(F)$ is arcwise connected to $f(0)\}$. It is easy to see that I_0 is a subinterval containing 0. Let $t_0 = \sup\{t \in I_0\}$. Suppose $t_0 < 1$. Then $f(t_0)$ is not contained in any U^j. In fact U^j is open and $U^j \cap T(F)$ is path-connected. Thus $f(t_0) \in U^j$ implies $f((t_0-\varepsilon, t_0+\varepsilon)) \subset U^j$ and $t_0 + \frac{\varepsilon}{2} \in I_0$. So $f(t_0) \in \Delta(L_0)$ for some limit separatrix. We can assume $f(I_0) \subset \delta(\Delta(L_0)) \cup \Delta(L_0)$, and $x_1 \in \delta'(\Delta(L_0))$ where $\delta(\Delta(L_0))$, $\delta'(\Delta(L_0))$ are the two open regions separated by $\Delta(L_0)$. From the discussion earlier, there is an arc passing through $\Delta(L_0) \cap T(F)$. Therefore, there is some $t_1 \in (t_0, 1]$ such that $f(t_1)$ is in some $\Delta(0)$ and $\Delta(0) \cap T(F)$ is arcwise connected to some $\Delta(0')$ containing $f(t_0-\varepsilon)$, $\varepsilon > 0$. Again we have $t_1 \in I_0$, a contradiction. Thus $t_0 = 1$ and $T(F)$ is path-connected. Q.E.D.

3.2.3. When (Ω, F) has limit separatrices, the union of all $N(0)$, 0 a canonical region, is not connected. See the example shown by Fig. 1.10.

Note that $T(F)$ is always contained in Ω. In general the closure of $T(F)$ in $\bar{\Omega}$ is not simply connected. In particular $T(F)$ may not be the closure of $\cup N(0)$, even not the closure of $\cup N(0)$ in

Ω. Study the following example.

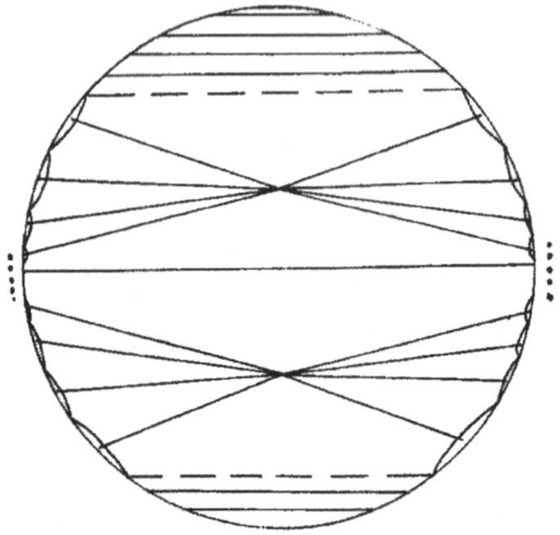

Fig. 3.3a

3.2.4. We define a metric d on $T(F)$ as follows. Let x, x' be a pair of points in $T(F)$. By Theorem 3.2.2, there is a unique arc L connecting x and x'. By 2.4.2, for every canonical region $N(O) \cap L$ is an arc if nonempty, and thus has a given length (3.1.3). We put

$$d(x,x') = \sum_{O \in CR} |N(O) \cap L| \quad . \tag{3.1}$$

The only nontrivial thing we need to verify in order to show that d is a metric is that $d(x,x') < \infty$. This is not obvious. For instance if we had chosen the center of a polygonoid P to be the barycenter of P, then the summation on the right hand side of (3.1) may diverge. See the example shown in Fig. 1.10.

The fact we proved in Theorem 3.2.2 that $T(F)$ is path-connected does not imply immediately that any two points in $T(F)$ are connected by an arc of finite length. See Fig. 3.3b, where the curve T is the image of a path, but the arc length between x_0 and x_1 is not

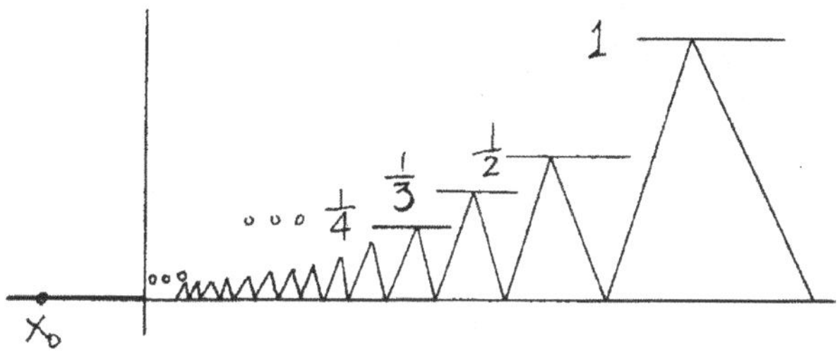

Fig. 3.3b

finite. However, we shall see that $(T(F),d)$ is not just "arcwise" connected, i.e. any two points in $T(F)$ is connected by an arc of finite length, the diameter of $(T(F),d)$ is also finite.

Lemma 3.2.5. *Assume the notation in 3.1.1. The sum of the lengths of* $\overline{x_p P_i}$ *and* $\overline{x_p P_j}$ *is less than twice the sum of the lengths of the two arcs on* $\partial\Omega$ *between* S_i *and* S_j *for any* i *and* j.

Proof. Denote $\overline{AB} = S_1$ and $\overline{CD} = S_2$. See Fig. 3.4. For two points on $\partial\Omega$, say E and F, we denote the length of the arc on $\partial\Omega$

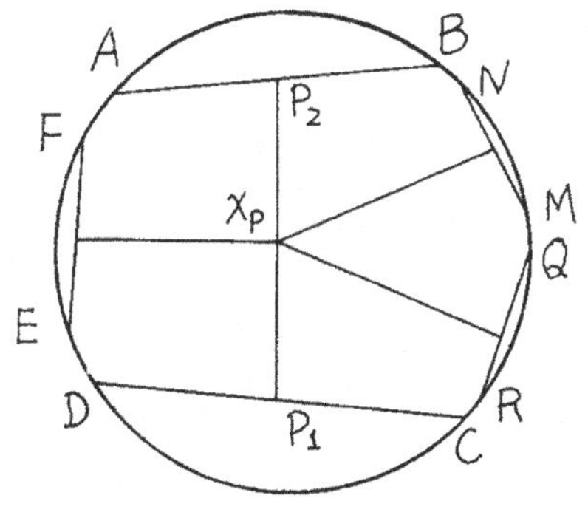

Fig. 3.4

measuring from E clockwise to F by $\widehat{EF}$ (so $\widehat{EF} + \widehat{FE} = 2\pi$).

Without loss of generality, we may assume that C, D, A, B are in clockwise order and thus $\widehat{DC} = \widehat{DA} + \widehat{AB} + \widehat{BC}$. When segments or circumference arcs appear in an equality, they stand for their lengths.

Case (1). The two sides S_i, S_j are $\overline{AB}$ and $\overline{CD}$. Then

$$(3.3)\qquad \overline{P_1x_p} + \overline{P_2x_p} = \frac{1}{2}(\overline{AB} + \overline{BC}) < \frac{1}{2}(\widehat{DA} + \widehat{BC}) .$$

The lemma is true.

Case (2). Only one side, say S_i, is $\overline{AB}$ or $\overline{CD}$. Let $S_j = \overline{EF}$. Then $\overline{EF} \le \overline{AB}$ and

$$(3.4)\qquad \overline{x_pP_j} \le \frac{1}{2}\widehat{EF} + \widehat{FA} + \frac{1}{2}\widehat{AB} + \overline{x_pP_2} .$$

Therefore

$$(3.5)\qquad \begin{aligned} \overline{x_pP_2} + \overline{x_pP_j} &\le 2\overline{x_pP_2} + \frac{1}{2}\widehat{AB} + \widehat{FA} + \frac{1}{2}\widehat{AB} \\ &= \widehat{DA} + \widehat{BC} + \widehat{AB} + \widehat{FA} \quad \text{(by (3.3))} \\ &\le \widehat{DE} + 2\widehat{FA} + 2\widehat{AB} + \widehat{BC} \\ &\le 2(\widehat{DE} + \widehat{FC}) \end{aligned}$$

This is the assertion in the lemma.

Case (3). Neither of the two sides S_i, S_j is $\overline{AB}$ or $\overline{CD}$, so that $\max\{S_i, S_j\} \le \min\{\overline{AB}, \overline{CD}\}$.

Case (3a). $S_j = \overline{EF}$, $S_i = \overline{QR}$ such that $EF \subset DA$, $QR \subset BC$.

We have

$$(3.6)\qquad \overline{x_pP_i} \le \frac{1}{2}\widehat{QR} + \widehat{RC} + \frac{1}{2}\widehat{CD} + \overline{x_pP_1}$$

Combining (3.4) and (3.6) we have again

$$\overline{x_pP_j} + \overline{x_pP_i} \leq \overline{P_1P_2} + \widehat{CD} + \widehat{RC} + \widehat{AB} + \widehat{FA}$$
$$\leq \widehat{DE} + 2\widehat{FA} + 2\widehat{AB} + \widehat{BO} + 2\widehat{RC} + 2\widehat{CD} \quad \text{(by (3.3))}$$
$$\leq 2(\widehat{RE} + \widehat{FQ})$$

Case (3b). $S_j = \overline{NM}$, $S_i = \overline{QR}$ such that both NM and QR are in BC. The discussion is similar to Case (3a). Q.E.D.

Theorem 3.2.6. *The function* $d: T(F) \times T(F) \to \mathbb{R}$ *defined in 3.2.4 is a metric.*

Proof. It is obvious that $d \geq 0$, and that $d(x_1,x_2) = 0$ implies $x_1 = x_2$. Now let $x_1, x_2, x_3 \in T(F)$. Let I_1, I_2, and I_3 be the unique arcs in $T(F)$ connecting x_1,x_2; x_2,x_3; and x_1,x_3, respectively. As in 2.4.3 we easily check $I_3 \subset I_1 \cup I_2$. It follows from (3.1) that the triangle inequality holds.

Now we come back to show that the sum on the right hand side is always finite. Actually we show that it is uniformly bounded: the *diameter of* $T(F)$

$$\text{(3.7)} \qquad \operatorname{diam}(T(F)) = \sup_{x_1,x_2 \in T(F)} \{d(x_1,x_2)\} \leq 4\pi$$

for any (Ω,F).

Let $x, x' \in T(F)$ such that $x \in \Delta(L)$ and $x' \in \Delta(L')$ for two chords $\Delta(L)$, $\Delta(L')$ in $K(F)$. We claim that $d(x,x')$ is less than two times the lengths of the portion of arcs between $\Delta(L)$ and $\Delta(L')$. The estimate (3.7) will follow from this quickly. By the additivity given by (3.1) we reduce the discussion to the case where both x, x' are in the same $\Delta(O)$ for a canonical region O. For $\Delta(O)$ of type (i) or (v) (3.1.2), it is obviously. For $\Delta(O)$ of one of types (ii), (iii) and (iv), this follows easily from Lemma 3.2.5. Q.E.D.

Naturally there arises a very interesting question. What is the least upper bound Λ of $\{\text{diam } T(F)\}$ for all possible $T(F)$? What is the significance of the real number Λ if it is not in the minimal algebraic field containing π, or containing e? Although this is not needed for the main goal of this paper, we give a partial answer and a heuristic proof.

Proposition 3.2.7. *The least upper bound* $\Lambda = 2\lambda$ *of* $\{\text{diam } T(F)\}$ *for all possible* $T(F)$ *is given by*

$$\lambda = \sum_{k=1}^{\infty}\left(\frac{1}{2^k \cos\frac{\pi}{2^{k+1}}} + \frac{1}{2^{2k-2}}\tan\frac{\pi}{2^{k+1}}\right) \sim 2.6476332557745\ldots \tag{3.8}$$

Proof. We claim that for the Kaplan diagram shown in Fig. 3.5,

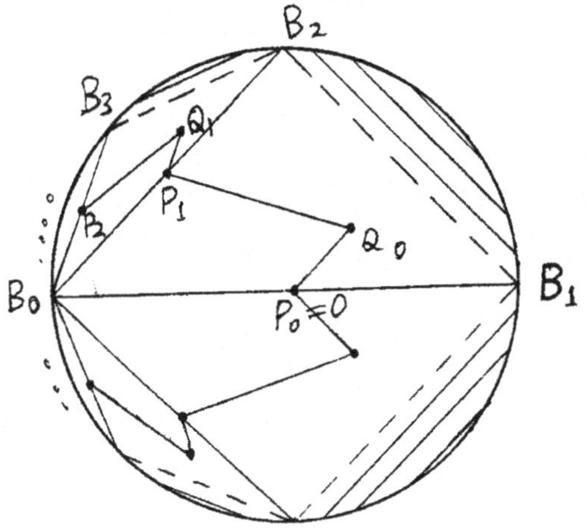

Fig. 3.5

$\text{diam}(T(F))$ attains the least upper bound Λ. Here $\overline{B_0B_1}$ is a diameter, each B_i bisects $\widehat{B_0B_{i-1}}$, $i = 2,3,\ldots$. Clearly $T(F)$ is simply reduced to an arc,

$$\text{(3.9)} \qquad \operatorname{diam}(T(F)) = 2\sum_{k=0}^{\infty} (\overline{P_kQ_k} + \overline{Q_kP_{k+1}}) .$$

By a little calculation

$$\text{(3.10)} \quad \overline{P_kQ_k} = 2^{-k}\left(\cos\frac{\pi}{2^{k+1}}\right)^{-1} , \qquad \overline{Q_kP_{k+1}} = 2^{-(2k-2)} \tan\frac{\pi}{2^{k+1}} .$$

Thus (3.8) follows from (3.9) and (3.10).

It is intuitively clear that a path Γ in $T(F)$ attaining the maximal length $\Lambda = 2\lambda$ should neither pass through any $N(0)$ with $\Delta(0)$ of type (i) or (v) (3.1.2), nor an open side of a polygonoid in $\Delta(0)$ of other types. In these cases, the arc length of the portion of Γ between two chords is less than one-half of the arc length on $\partial\Omega$ between the two chords. Also both $T(F)$ and Γ have to be symmetric either with respect to the center 0 or to an axis B_0B_1. We may assume the latter is true (flip over the lower half disc in Fig. 3.4, to get the former case).

To maximize the ratio of $\operatorname{diam}(N(p))$ over the corresponding arc lengths on $\partial\Omega$, one concludes that the best strategy is to have the third longest side of P equal to the second, and let it be as long as possible; thus P should be a triangle. Q.E.D.

Corollary 3.2.8. *For any foliation* (Ω,F), *the associated metric space* $(T(F),d)$ *is an* $\mathbb{R}$*-tree.*

Proof. It follows easily from Theorems 3.2.2 and 3.2.6. Q.E.D.

We want the $\mathbb{R}$-tree $(T(F),d)$ associated to a foliation (Ω,F) to be equipped with a certain extra structure so that it contains exactly all the information for the C*-algebra $C^*(\Omega,F)$.

Observe the five types of canonical regions (Fig. 3.2). Each of them has exactly one center except that of type (iv) which has two. These centers will be called vertices of the $\mathbb{R}$-tree $(T(F),d)$. They shall represent the "cells" of the noncommutative CW-complex $C^*(\Omega,F)$. To classify these cells, notice that the chords in the image $\Delta(0)$ of a canonical region 0 may intersect $\partial\Omega$ on one (types (i), (ii)) or two (types (iii), (iv) and (v)) arcs. When considering it as a foliated sphere with one singularity, a canonical region of the former types is a rose petal and that of the latter types is a strip encircling more complicated regions. We will distinguish the centers of these two types of canonical regions by a $\mathbb{Z}_2$-grading with those of types (i) and (ii) of degree 1 and the others of degree 0.

After "straightening out" the leaves, the canonical regions with their boundary corresponding to vertices of deg 1 and deg 0 are illustrated respectively in Fig. 3.6 (a) and (b). There, F is a measure zero closed subset of $\mathbb{R}$ (which can be empty) (cf. 4.2.8 and the beginning of §4.1).

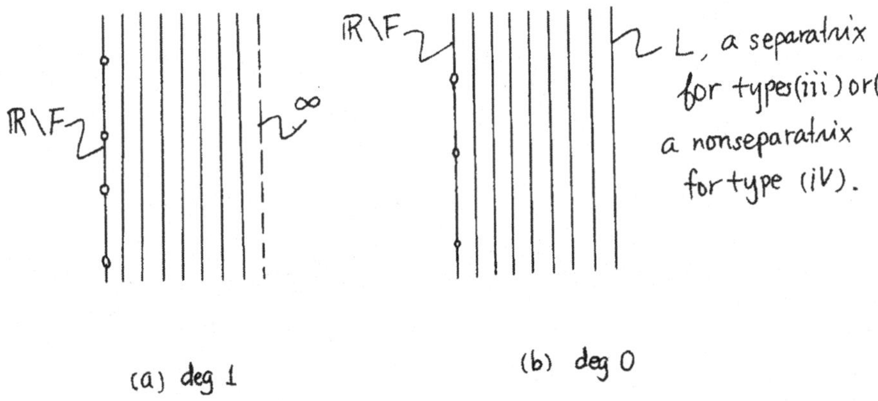

Fig. 3.6

Definition 3.2.9. We equip the $\mathbb{R}$-tree $(T(F),d)$ with some extra structure as follows. Define a subset V of $T(F)$ by (recall 3.1.2)

$$V = \{x \in T(F) \mid x \text{ is a center of } \Delta(0) \text{ for some canonical region } 0\}$$

Moreover, we $\mathbf{Z}_2$-grade the set V as follows: When x is a center of $\Delta(0)$ which is of type (i) or (ii), we decree that x has $\mathbf{Z}_2$-degree 1, i.e. $\deg x = 1$; for all the remaining $x \in V$, we decree that x has $\mathbf{Z}_2$-degree 0, i.e. $\deg x = 0$. Denote $V_i = \{x \in V \mid \deg x = i\}$, $i = 0,1$.

Proposition 3.2.10. *Let* (Ω, F) *be a foliation. Then*

(1) *The set* V *is a countable discrete subset of* $T(F)$.

(2) *Both the set of branched points* B_T *and the set of pending points* P_T *are contained in* V *(see 2.3.2).*

(3) *For each* $x \in P_T$, $\deg x = 1$.

(4) *For each* $x \in T(F)$, *every component of* $T(F)\setminus\{x\}$ *contains at least one point of* V.

Proof. (1) and (4) are obvious by our construction.

Every branch point x in $T(F)$ has to be the center of a polygonoid P, which must be contained in some $\Delta(0)$ of types (ii)-(iv). A pending point x has to be the center of an abutment, i.e., some $\Delta(0)$ of type (i). Thus (2) and (3) hold. Q.E.D.

We notice that by our construction a point x in $V - B_T \cup P_T$ is either the center of some $\Delta(0)$ of type (v), or the center of some $\Delta(0)$ of type (ii) containing a polygonoid with exactly two closed sides. These two cases are distinguished by the $\mathbf{Z}_2$-grading.

Theorem 3.2.11. *Let* (Ω,F) *be a foliation. Let* $(T(F),d,V)$ *be defined as above. Then* $(T(F),d,V,E)$ *is a regular* $\mathbb{R}$*-tree (Definition 2.3.3).*

Proof. This follows from Proposition 3.2.10 and Corollary 3.2.8. Q.E.D.

3.2.12 Let E be the set of edges of $(T(F),d,V)$ as defined as 2.3.2. Recall (Lemma 2.3.4) that there is an injective map $\gamma\colon E \longrightarrow V\times\bar{V}$, $\gamma(y) = (x,x_y)$, for $y \in E_x$ and that $I\cap\bar{V} = \{x,x_y\}$ for the unique oriented closed segment $\vec{I}$ from x to x_y.

Note that $B_T\cap V_0$ is the set of vertices which are the centers of the various $\Delta(0)$ of types (iii) and (iv) (Fig. 3.1). Let $x \in B_T\cap V_0$, which is then the center of a polygonoid P with an open side, say S_k. The oriented segment $\overrightarrow{xp_k} \in F_x$, where P_k is the middle point of S_k, represents a unique $y_k \in E_x$. We introduce a $\mathbb{Z}_2$-grading on E_x by letting $\deg y_k = 1$ for this $y_k \in E_x$ and $\deg y = 0$ for the remaining $y \in E_x$. Thus there is exactly one edge $y \in E_x$ with $\deg y = 1$ for each $x \in B_t\cap V_0$ which is exactly on the side consisting of a single leaf in the boundary of the region shown in Fig. 3.6b. We define $\deg y = 0$, $y \in E_x$ for all other types of $x \in V$. For convenience in illustrations we shall put an arrow on the oriented closed segment from x_y to x pointing to x, if $\deg y = 1$, where $\gamma(y) = (x,x_y)$, and we shall say "there is an arrow coming from x_y to x". A vertex of degree 0 will be denoted by a small circle while a vertex of degree 1 by a black solid dot.

Clearly both deg $y = 1$ and deg $\bar{y} = 1$ ($\bar{y}$ is the inverse of y, Lemma 2.3.4) happen for $y \in E_x$ iff x and x_y are the centers of some $\Delta(0)$ of type (iv).

Definition 3.2.13. Let (Ω, F) be a foliation. The *associated tree of* (Ω, F) is the regular $\mathbb{R}$-tree $T(F) = (T(F), d, V, E)$, where the set of vertices V is $\mathbb{Z}_2$-graded and for each $x \in B_T \cap V_0$, the set E_x is $\mathbb{Z}_2$-graded as defined above.

Denote $V_{0,2} = \{x \in V_0 \mid \mathrm{inc}(x) = 2\}$, the center of $\Delta(0)$ of type (v). The tree associated to (Ω, F) will be denoted by $T(F)$.

Proposition 3.2.14. *The following are true for* $T(F)$.

(1) *Suppose* $x \in P_T$. *Let* $x_y \in \bar{V}$ *corresponding to the unique* $y \in E_x$. *Then it cannot happen that* $x_y \in B_T \cap V_0$ *with* deg $\bar{y} = 1$, *i.e. an arrow does not come from a pending point.*

(2) *An arrow does not come from any* $x \in V_{0,2}$.

(3) *Assume* $x \in P_T \cap V_{0,2}$. *Then for each* $y \in E_x$, *we have* $x_y \notin P_T \cup V_{0,2}$. (In short we shall say that the set $P_T \cup V_{0,2}$ is *disconnected with respect to* E.)

In illustration, Proposition 3.2.14 asserts that the following diagrams cannot be contained in $T(F)$:

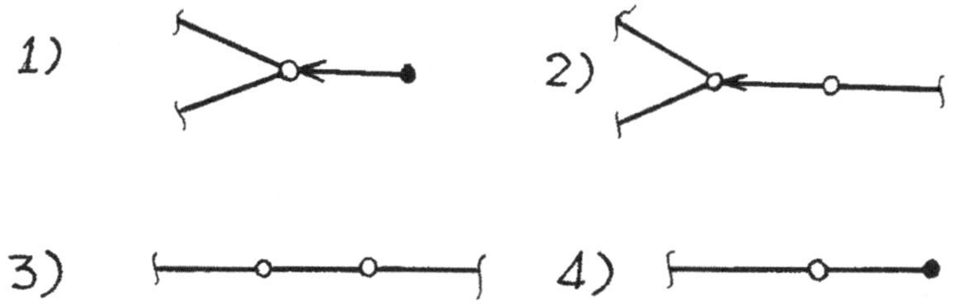

Fig. 3.7

More intuitively, we have

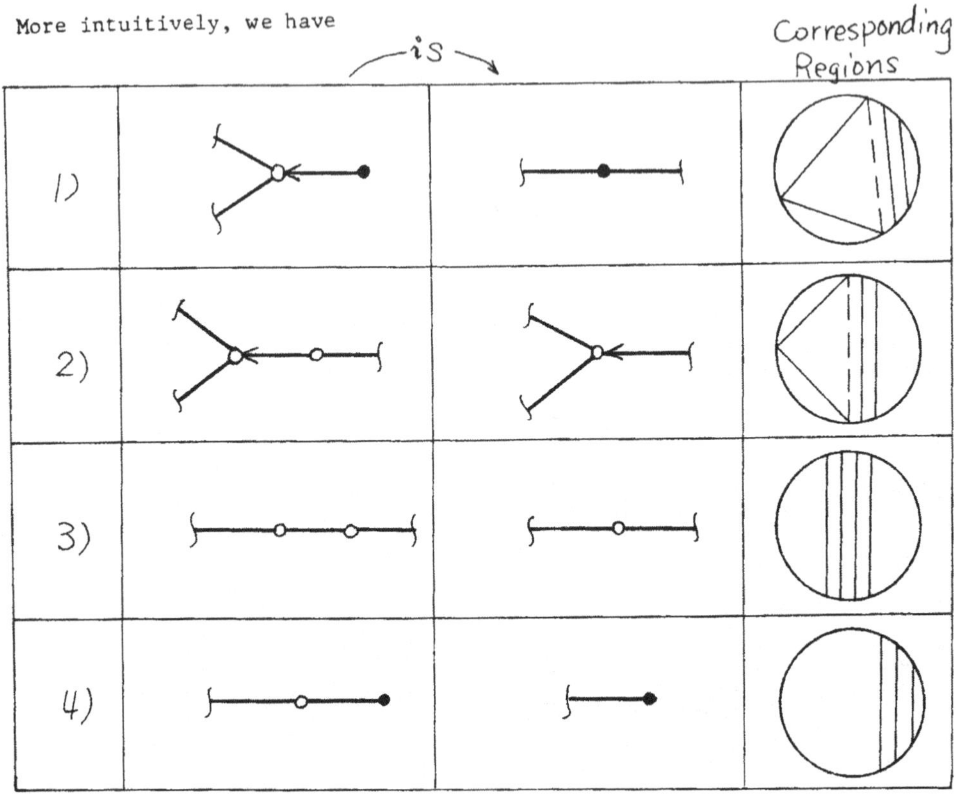

Fig. 3.8

Proof of Proposition 3.2.14. For (1) suppose that $x_y \in B_T \cap V_0$ and $\deg \bar{y} = 1$. Then the unique closed segment $\bar{\mu}$ connecting x_y and x would intersect the open side of the corresponding polygonoid. Since $\bar{\mu} \cap V = \{x, x_y\}$, there is no separatrix L such that $\bar{\mu} \cap \Delta(L) \neq \emptyset$. This contradicts the assumption that x is the center of an abutment.

Now assume $x \in V_{0,2}$. This is equivalent to saying that x is the center of some $\Delta(0)$ of type (v). Suppose $x_y \in B_T \cap V_0$ and $\deg \bar{y} = 1$. As in (1), we have $\bar{\mu} \cap V = \{x, x_y\}$. Let $\Delta(L)$ be one of two chords contained in the boundary of $\Delta(0)$ which intersects with $\bar{\mu}$. Then L is a separatrix but $\Delta(L)$ does not satisfy the condition (1) in Theorem 1.5.7. On the other hand, $\Delta(L)$ does not satisfy the condition (2) there either, because otherwise $\Delta(L) \cap T(F)$ would be the point x_y. This is a contradiction.

For the same reason if $x \in V_{0,2} \cup P_T$, for any $y \in E_x$, we have $x_y \notin V_{0,2}$ and $x_y \notin P_T$. This is (3). Q.E.D.

Proposition 3.2.15. $(\bar{V}_1 \setminus V_1) \not\subset T(F)$. *Equivalently,* $(\bar{V}_1 \setminus V_1) \subset \partial\Omega$.

Proof. We notice that any element in V_1 is either in P_T or the center of some $\Delta(0)$ of type (ii).

Suppose $\{x_n\} \subset \bar{V}_1$ such that $x_n \to x$.

First assume that there is an infinite subsequence $\{x_{n_i}\} \subset P_T$. Then each x_{n_i} is the center of an abutment Q_i in $\bar{\Omega}$. Since $\operatorname{diam} Q_i \to 0$, so $x = \lim_i x_{n_i} \in \partial\bar{\Omega}$.

Now assume $\{x_n\} \cap P_T = \emptyset$ after throwing out finitely many points. Each point x_n is the center of some $\Delta(0_n)$ of type (ii). It follows that all but finitely many x_n's are centers of acute polygonoids. By Theorem 1.4.4, $\operatorname{diam}(\Delta(0_i)) \to 0$. Thus $x \in \partial\Omega$.

Q.E.D.

Lemma 3.2.16. *Let* (T,d,V,E) *be a regular* $\mathbb{R}$*-tree. Suppose* $x \in (\bar{V} \setminus V) \cap T$. *Then* $\operatorname{inc}(x) = 2$.

Proof. Since $B_T \cup P_T \subset V$, it follows that x is not in $B_T \cup P_T$. Since $x \in T$, it follows that $\operatorname{inc}(x) = 2$. Q.E.D.

Example 3.2.17. Let T be the subspace in $\mathbb{R}^2$: $T = [-1,1] \times \{0\} \cup \{1, \frac{1}{2}, \ldots\} \times [0,1]$. Define an obvious metric d to make T an $\mathbb{R}$-tree. Let $V = B_T \cup P_T$. Let $x_n = (\frac{1}{n}, 0)$, $n = 1,2,\ldots$. Then $\operatorname{inc}(x_n) = 3$, $x_n \to 0$, but $\operatorname{inc} 0 = 2$.

We can easily construct a similar example even with $\mathrm{inc}(x_n) = \infty$ for all $n = 1,2,\ldots$ but $\mathrm{inc}(x_0) = 2$.

Proposition 3.2.18. (1) *For every* $z \in V'$ *(Definition 2.3.3), there is an* $\ell_z > 0$ *and a closed segment* $\Gamma_z \subset T(F)$ *containing* z *in its interior such that* (i) $\bar{V} \cap B(z,\ell_z) \subset \Gamma_z$, *where* $B(z,\ell_z)$ *is the ball in* $T(F)$; (ii) *for all* $x \in B(z,\ell_z) \cap B_{T(F)} \cap V_0$, *the segment* Γ_z *contains the arrow pointing to* x, *i.e. for some* $y \in E_x$, $\deg y = 1$, $I_{x,x_y} \cap \Gamma_z \subset I_{x,x'}$ *for some* $x' \in \mathring{I}_{x,x_y}$ *(see Proposition 2.2.8)*; (iii) $\Gamma_z \cap V_1 = \emptyset$.

(2) *Such a collection* $\{\Gamma_z\}_{z\in V'}$ *can be fixed so that for any* $z, z' \in V'$, *the two arcs* Γ_z *and* $\Gamma_{z'}$ *are either disjoint or coincide.*

Proof. (1) There is a limit separatrix L_z such that $\Delta(L_z) \cap T(F) = \{z\}$. Without loss of generality, we may assume that there is a sequence $\{x_n\} \subset V$ converging to z from both sides. We simply choose Γ_z to be a closed segment contained in the arc T_z constructed in the proof of Theorem 3.2.2, and which still has z in its interior.

There exist two separatrices L_1, L_2 such that (see Fig. 3.8)

(1) $\Delta(L_z)$ separates $\Delta(L_1)$ and $\Delta(L_2)$;

(2) $\Delta(L_i) \cap T(F) = \{z_i\} \subset \Gamma_z$, $i = 1,2$;

(3) Let $\Delta(L_i)$ intersect $\partial\Omega$ at A_i, B_i (in appropriate order), $i = 1,2$. Then $d_\Omega(z,\overline{A_1A_2}) > \frac{1}{4}$, $d_\Omega(z,\overline{B_1B_2}) > \frac{1}{4}$.

Now let $\ell_z = \min\{d_\Omega(z,z_i),\frac{1}{4}\} > 0$, and let U_z be the region in Ω between $\Delta(L_1)$ and $\Delta(L_2)$.

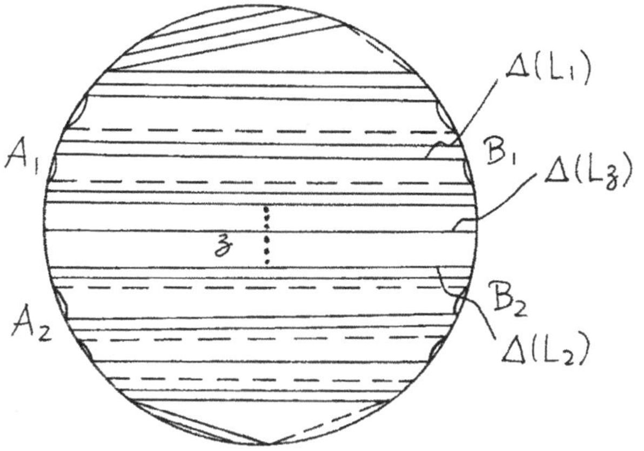

Fig. 3.8

Suppose that $x \in B(z, \ell_z) \cap \bar{V}$ and $x \notin \Gamma_z$. Then x is either (a) the middle point of some $\Delta(L)$, for a limit separatrix $L \subset U_z$; or (b) a center of some $\Delta(0)$ for a canonical region 0 such that $\Delta(0) \subset U_z$. However since $x \notin \Gamma_z$, in either case L or $\Delta(0)$ would be separated apart from z by either $\overline{A_1A_2}$ or $\overline{B_1B_2}$. Recalling the construction of the path f in the proof of the theorem, we see that in fact the unions of all the L's with middle points on Γ_z and the $\Delta(0)$'s with centers on Γ_z have already covered up the quadrangle $\square A_1B_1B_2A_2$. It follows that $d_{T(F)}(x,z) > d_{\Omega}(x,z) > \frac{1}{4} > \ell_z$, a contradiction.

By the construction of Γ_z, all the polygonoids intersecting Γ_z are obtuse and the middle points of all their open sides, except possibly the two closest to the ends of Γ_z, are contained in Γ_z. Thus (ii) can be satisfied. By Proposition 3.2.15, we may shorten Γ_z a little to satisfy (iii).

(2) We shall call a closed segment Γ admissible if both the extremities of Γ are not in V' and for any $z \in \Gamma \cap V'$, there is some $\ell_z > 0$ such that (1) holds. By the discussion above, every $z \in V'$ is contained in the interior of an admissible segment. It is easy to see that the set $\bar{V}$ is σ-compact (Proposition 3.2.20). Thus there is a sequence $\{\Gamma_n\}$ of admissible segments such that $\bigcup_{n=1}^{\infty} \Gamma_n \supset V'$. We construct a new sequence $\{\Gamma'_m\}$ of mutual disjoint admissible segments from $\{\Gamma_n\}$ as follows.

(a) Let $\Gamma'_1 = \Gamma_1$.

(b) Suppose that the mutual disjoint admissible segments $\Gamma'_1,\dots,\Gamma'_{n_k}$ have been constructed from $\Gamma_1,\dots,\Gamma_k$ such that $\bigcup_{j=1}^{n_k} \Gamma'_j \cap V' = \bigcup_{i=1}^{k} \Gamma_i \cap V'$. We consider $\Gamma_{k+1} \setminus \bigcup_{j=1}^{n_k} \Gamma'_j$, which is a union of at most (n_k+1)-segments, $\Gamma_{k+1,1},\dots,\Gamma_{k+1,m}$, all of them are open except for two (if $m \geq 2$) which are half open. Suppose $m \geq 2$ and fix $1 \leq i \leq m$. Let x be an extremity at $\Gamma_{k+1,i}$, i.e., $x \in \bar{\Gamma}_{k+1,i} \setminus \Gamma_{k+1,i}$. Then x is either in $B_T \cap V_0$ or an extremity of some Γ_i. In both cases $x \notin V'$. Therefore $\Gamma_{k+1,i}$ may be shortened a bit to get an admissible $\Gamma'_{k+1,1}$ such that $\Gamma_{k+1,i} \cap V' = \Gamma'_{k+1,1} \cap V'$. We denote $\{\Gamma'_1,\dots,\Gamma'_{n_k},\Gamma'_{k+1,1},\dots,\Gamma'_{k+1,m}\}$ by $\{\Gamma'_1,\dots,\Gamma'_{n_{k+1}}\}$. Clearly $\{\Gamma'_1,\dots,\Gamma'_{n_{k+1}}\}$ is again a finite mutual disjoint collection of admissible segments and $\bigcup_{j=1}^{n_{k+1}} \Gamma'_j \cap V' = \bigcup_{i=1}^{k+1} \Gamma_i \cap V'$. Thus $V' \subset \bigcup_{j=1}^{\infty} \Gamma'_j$ and for each $z \in V'$, there is a unique $\Gamma'_j \ni z$ and we let $\Gamma_z = \Gamma'_j$. Q.E.D.

Remark 3.2.19. We shall assume that such a collection $\{\Gamma_z\}$ has been chosen once and for all. For every $x \in V$, there is at most one edge $y \in E_x$ of deg 0 contained in some Γ_z, $z \in V'$. We shall always label all the elements of deg 0 in E_x by $y_1,\ldots,y_n$, $2 \leq n \leq \infty$, in such a way that only y_1 is possibly contained in some Γ_z, $z \in V'$.

Proposition 3.2.20. *Let* $x \in T(F)$ *and let* $T_i(x)$ *be a component of* $T(F)\setminus\{x\}$. *If* $T_i(x) \cup \{x\}$ *is noncompact, then it contains infinitely many vertices.*

Proof. For all types of $\Delta(0)$, the nerve is compact and contains only one or two vertices. If $T_i(x) \cup \{x\}$ contains only finitely many vertices, then it contains finitely many $N(0)$'s and does not intersect with V'. Thus $T_i(x) \cup \{x\}$ is the union of these finitely many $N(0)$'s with a compact subset contained in a certain $N(0_x)$. Hence it is compact. Q.E.D.

In particular, an "open end" of $T(F)$ always contains infinitely many vertices in any neighborhood of it. On the other hand, for a "closed end", i.e., a pending point, there is of course a neighborhood containing no vertices except itself.

3.2.21 Note that a tree T associated to a foliation may not be locally compact as a metric space. Consider (Ω, F) given by Fig. 1.8b. Its associated tree is the one shown in Fig. 3.9. It is

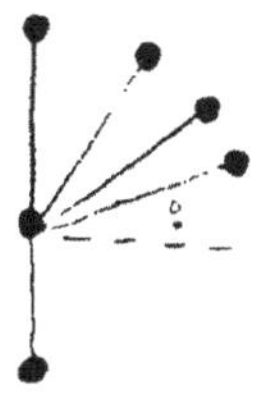

Fig. 3.9

clear that the point 0 does not have a compact neighborhood.

As Proposition 3.2.15 and Proposition 3.2.18 indicate, however, the closure $\bar{V}$ of the vertices in T is locally compact.

Proposition 3.2.22. *As a metric space, the tree* T(F) *associated to a foliation* (Ω,F) *is always* σ*-compact. In particular, the set* $\bar{V}$ *is always* σ*-compact.*

Proof. Recall that T(F) is a countable union of the $N(0)$ and some points (3.2.1), each of them is σ-compact. Thus T(F) is σ-compact. Q.E.D.

Remark 3.2.23. The induced topology on T(F) as a subset of Ω is in general different from the topology given by the metric d. For example, the tree T(F) (Fig. 3.10) associated to the (Ω,F)

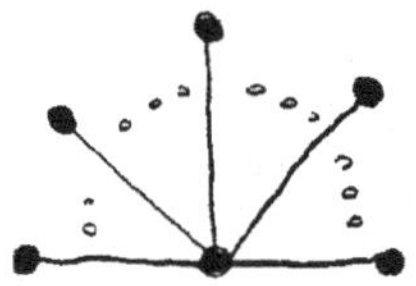

Fig. 3.10

shown in Fig. 1.8a is homeomorphic to the one shown in Fig. 3.9, as $\mathbb{R}$-trees, but they are not homeomorphic as subsets of (Ω,d_Ω). In fact, these two associated trees are "similar" (Definition 3.3.3) (and thus the C*-algebras of these two foliations are isomorphic, while the two foliations are, of course, not conjugate). Therefore the steps we spent in §3.1 to define a metric to make T(F) an

R-tree are not vacuous.

Question. Show an $\mathbb{R}$-tree is σ-compact iff it is separable.

3.3 **An intrinsic definition of distinguished trees**

Some of the properties of the trees associated to a foliation which we discussed in 3.2 can be used to characterize among all regular $\mathbb{R}$-trees with vertices and edges $\mathbf{Z}_2$-graded. These come from foliations of the plane. We summarize these properties with axioms for distinguished trees. The reader should consult 3.2 for the geometric motivations for these axioms.

Definition 3.3.1. A *distinguished tree* $T = (T,d,V,E)$ is a regular $\mathbb{R}$-tree with $\text{Card}(V) \geq 3$ such that the following axioms are satisfied.

(1) The set of vertices V is $\mathbf{Z}_2$-graded: $V = V_0 \cup V_1$ such that $P_T \subset V_1$.

(2) The set E of edges is $\mathbf{Z}_2$-graded such that for each $x \in B_T \cap V_0$ there is exactly one $y \in E_x$ with $\deg y = 1$. If $y \in E_x$, $x \notin B_T \cap V_1$, then $\deg y = 0$.

Assume the terminology in 3.2.12, Propositions 3.2.14 and 3.2.18.

(3) An arrow does not come from a vertex in $P_T \cup V_{0,2}$.

(4) The set $P_T \cup V_{0,2}$ is disconnected with respect to E.

(5) For every $z \in V'$, there is an open arc $\Gamma_z \subset T$ and a $\ell_z > 0$ such that (i) $B(z,\ell_z) \cap V \subset \Gamma_z$; (ii) Γ_z contains the arrow point to x, for each $x \in B(z,\ell_z) \cap V_0 \cap B_T$; (iii) $\Gamma_z \cap V_1 = \emptyset$.

Remark 3.3.2. Since (2), Proposition 3.2.18 follows from (1), Proposition 3.2.18, which is axiom (5) above, we shall assume the choice made in Remark 3.2.19. More precisely we assume

(1) There is a fixed collection $\{\Gamma_z\}_{z \in V'}$ of closed segments in T such that for each $z \in V'$, the segment Γ_z satisfies (1), Proposition 3.2.18, and for any $z, z' \in V'$, the two segments Γ_z and $\Gamma_{z'}$ are either disjoint or coincide.

(2) If $x \in V_0$ then E_x is enumerated as $\{y, y_1, y_2, \ldots\}$ where $\deg y = 1$ if $x \in V_0 \cap B_T$ and only y and y_1 are possibly contained in some Γ_z.

(3) If $x \in V_1$, then E_x can be enumerated as $\{y_1, y_2, \ldots\}$, where only y_1 can possibly be contained in some Γ_z,

Definition 3.3.3. Two distinguished trees $T = (T,d,V,E)$ and $T' = (T',d',V',E')$ are *similar* if there is a homeomorphism from (T,d) onto (T',d') which defines an isomorphism of the corresponding regular $\mathbb{R}$-trees (2.3.6), and furthermore preserves the $\mathbf{Z}_2$-gradings on V and E.

Theorem 3.3.4. *Let* (Ω,F) *be a foliation with* T_2*-separatrices. Then the tree* $T(F)$ *associated to* (Ω,F) *is a distinguished tree.*

Proof. This follows directly from Theorem 3.2.11, Proposition 3.2.14, Proposition 3.2.18 and Proposition 3.2.22. Q.E.D.

Proposition 3.3.5. *Suppose that two foliations* (Ω,F) *and* (Ω',F') *with* T_2*-separatrices are conjugate. Then the corresponding associated distinguished trees* $T(F)$ *and* $T(F')$ *are similar.*

Proof. By Theorem 1.4.9 the Kaplan diagrams of (Ω,F) and (Ω,F') are similar. We can extend the homeomorphism on $\partial\Omega$ onto $\bar{\Omega}$. Q.E.D.

Examples 3.3.6. (1) Suppose Card V = 3. Then there is only one distinguished tree up to isomorphis, as shown in Fig. 3.11. It is

Fig. 3.11

the associated tree of the foliation shown in Fig. 1.3.

(2) Suppose Card V = 4. There are exactly two non-isomorphic distinguished trees (Fig. 3.12). Notice that Fig. 3.12(a) is the associated tree for the two non-conjugate foliations shown in Fig. 3.13(a1) and Fig. 3.13(a2). Fig. 3.12(b) is the associated tree for Fig. 3.13(b). Note that Fig. 3.13(a2) represents the foliation of two non-conjugate flows. In fact the two flows in Fig. 3.13(c) are not conjugate. Thus there are four non-conjugate flows with four canonical regions.

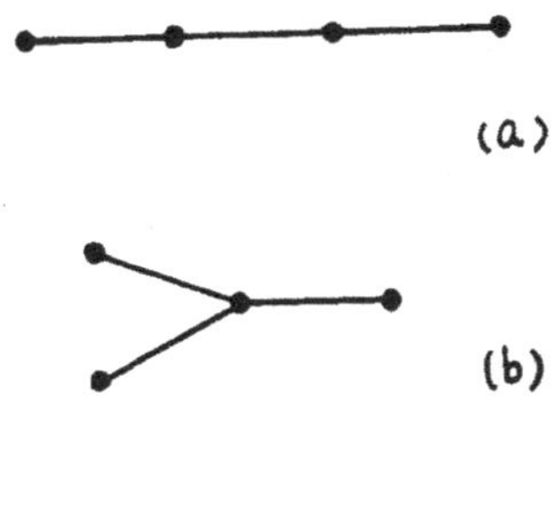

Fig. 3.12

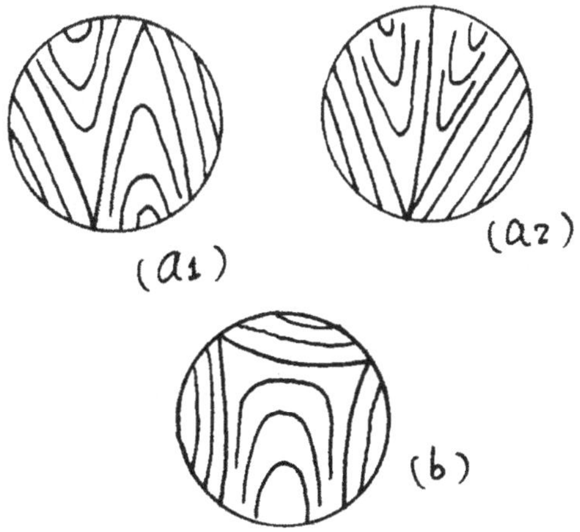

Fig. 3.13

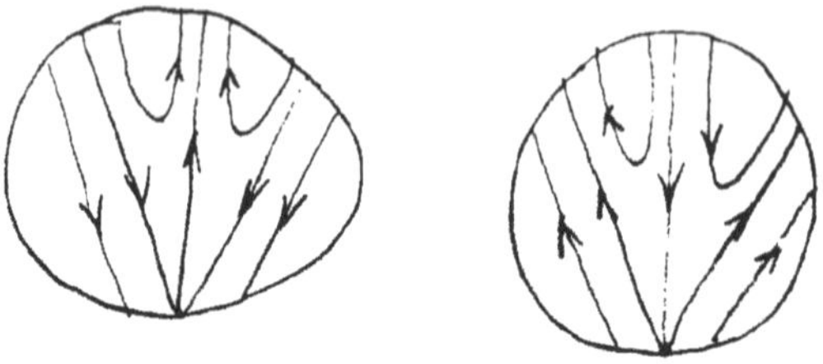

Fig. 3.13(c)

(3) Now assume Card = 5. There are altogether five distinguished trees up to isomorphism (see Fig. 3.14). The corresponding Kaplan diagrams are shown in Fig. 3.15.

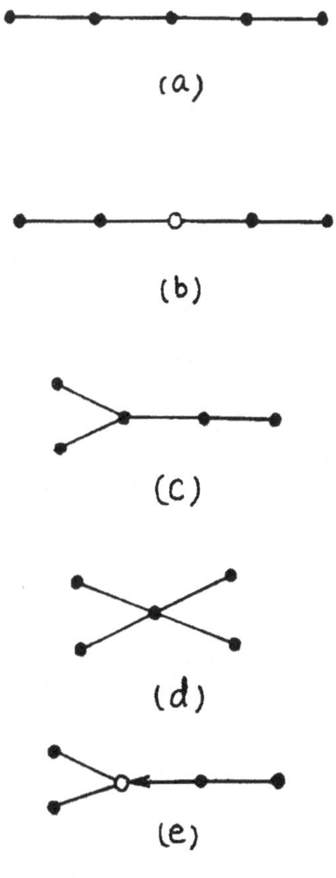

Fig. 3.14

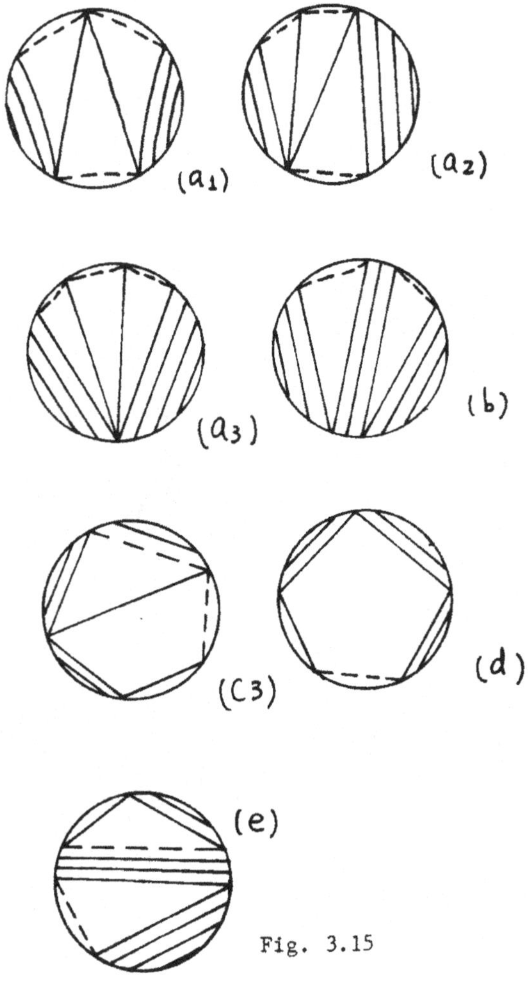

Fig. 3.15

Note that there are three non-conjugate Kaplan diagrams for the tree shown in Fig. 3.14(c). One of them is Fig. 3.15(c3); the other two are given by Fig. 1.7(a),(b).

Theorem 3.3.7. *Let* (T,d,V,E) *be a distinguished tree such that* V *is closed. Then there is a (not unique) foliation* $(\Omega,\mathcal{F})$ *without limit separatrices such that* (T,d,V,E) *is similar to the associated tree of* $(\Omega,\mathcal{F})$. *Conversely, if a foliation* $(\Omega,\mathcal{F})$ *has no limit separatrices, then the associated tree* $T(\mathcal{F}) = (T(\mathcal{F}),d,V,E)$ *is a distinguished tree such that* V *is closed.*

Remark 3.3.8. Later (Theorem 3.5.6), we shall remove the restriction that V is closed, or equivalently that the foliation has no limit separatrices. (The discussion in Markus' paper [Mar] is restricted to this special case.) Although the following proof will not be used in the general situation, we give the proof for this important special case because it is simple and informative.

Proof. The second part of the theorem is a direct corollary of Theorem 3.2.11, Proposition 3.2.14 and Proposition 3.2.22. From the construction of $T(\mathcal{F})$ we see that the absence of limit separatrices implies that V is closed.

Let us now consider the first part of the theorem. Let T = (T,d,V,E) be a distinguished tree such that V is closed. By Theorem 2.3.5, T is an ordinary tree. Thus we shall assume T is a $\mathbf{Z}$-tree, i.e., the distance between any two adjacent vertices is 1.

First we prove by induction that if $\operatorname{diam}(T) < \infty$, then there exists a Kaplan diagram $K(\mathcal{F})$ such that $T(\mathcal{F})$ is similar to T.

Clearly we always have $\operatorname{diam}(T) \geq 2$. Assume $\operatorname{diam}(T) = 2$. Then $T(F)$ has only one point x_0 with $\operatorname{Inc}(x_0) \geq 2$ (see Fig. 3.16). Such $K(F)$ exists (see Fig. 1.9).

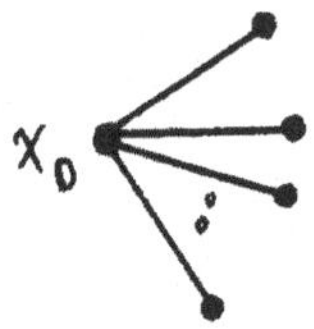

Fig. 3.16

Suppose that we have shown for any distinguished tree T with $2 \leq \operatorname{diam}(T) \leq n$ that there exists a Kaplan diagram $K(F)$ such that $T(F)$ is similar to T. Let $T_1 = (V^1, E^1)$ be a distinguished tree with $\operatorname{diam} T_1 = n+1$. Observe first that if n is even, then T_1 has a unique "central edge" y_0, such that for each extremity x_0 of y_0, $\max\{d(x,x_0) \mid x \in V^1\} = \frac{n}{2}+1$. (Consider the "central edge" of a path of length $(n+1)$ in T_1. If there are two such paths, their central edges coincide because otherwise it is easily seen that there would exist a path of length $\geq n+2$ in T_1.) If n is odd then T_1 has a unique "central vertex" x_0 such that $\max\{d(x,x_0) \mid x \in V^1\} = \frac{n+1}{2}$. We discuss these two cases separately.

First suppose n is even. Fix an extremity x_0 of y_0. Then the set $V_b = \{x \in V^1 \mid d(x,x_0) = \frac{n}{2}+1\}$ is contained in P_{T_1}.

Let $x \in V_b$. There is a unique $x_1 \in V^1$ adjacent to x. Consider the path connecting x_0 and x_1 of length $\frac{n}{2}$. We see that there is also a unique $x_2 \in V^1 \backslash P_{T_1}$ adjacent to x_1. According to the gradings of x_1, x_2 and possibly of the edges $[x_1,x_2]$, $[x_2,x_1]$, we classify into the following four situations and operate accordingly.

(i) $\deg x_1 = \deg x_2 = 1$, or $\deg x_1 = 1$, $x_2 \in V_0 \cap B_{T_1}$, but $\deg[x_1,x_2] = \deg[x_2,x_1] = 0$.

We remove all the $x \in P_{T_1}$ adjacent to x_1. This changes x_1 to a pending point, as shown in Figs. 3.17(a) and (b).

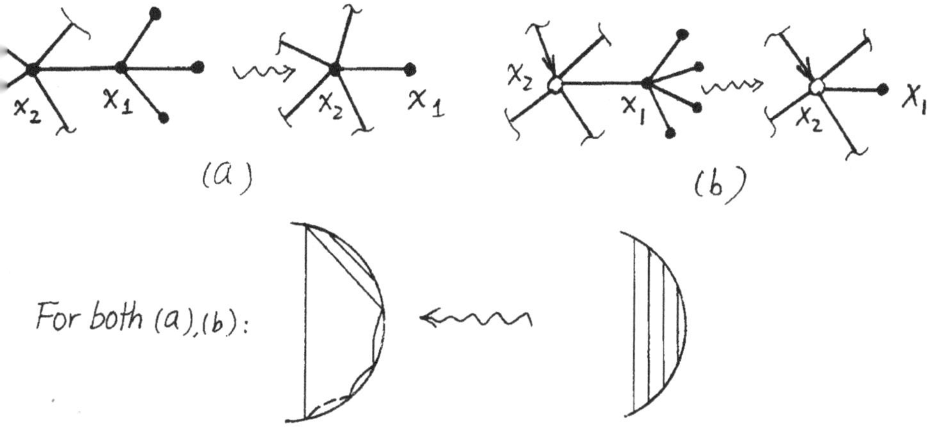

Fig. 3.17

(ii) $\deg x_1 = 1$, $x_2 \in V_0 \cap B_T$ and $\deg[x_1,x_2] = 1$. We remove x_1 together with all the pending vertices adjacent to it. Then we change $\deg x_2$ from 0 to 1. See Fig. 3.18(a).

(iii) $\deg x_1 = 1$, $\deg x_2 = 0$ but $x_2 \notin B_T$. We remove only x_2. See Fig. 3.19(a).

(iv) $\deg x_1 = 0$, $\deg x_2 = 0$ or 1.

Then of course $\deg[x_2,x_1] = 1$. We remove all the pending points adjacent to x_1 and change deg of x_1 to 1. See Fig. 3.20(a).

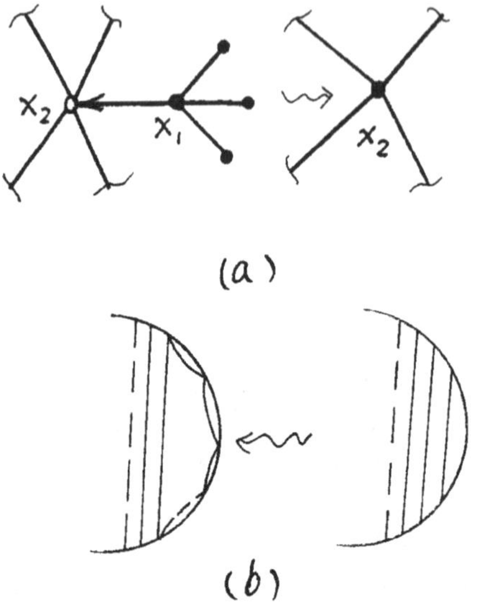

Fig. 3.18

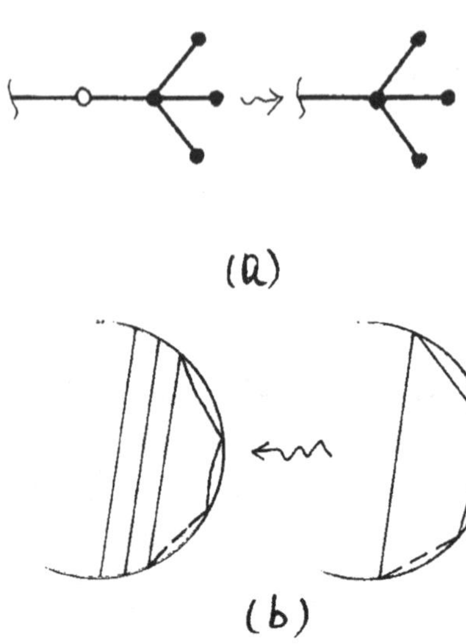

Fig. 3.19

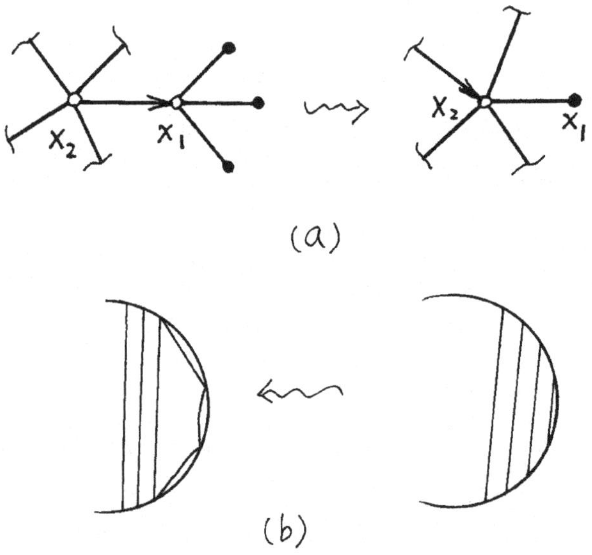

Fig. 3.20

After completing the above operations on T_1, we obtain a distinguished tree T with $diam(T) = n$. By the induction hypothesis there is a Kaplan diagram $K(F)$ such that the associated tree of $K(F)$ is similar to T.

Now we show that for all the four situations, the inverse operation on $K(F)$ recovers the Kaplan diagram $K(F_1)$ whose associated tree is similar to T_1. See Figs. 3.17(c), 3.18(b), 3.19(b) and 3.20(b). They are self-explanatory.

If $diam(T_1) = n+1$ and n is odd, we replace an extremity x_0 of the central edge in the n even case by the central vertex x_0 of T_1. The proof is the same. Our induction is complete.

Now let $T = (T,d,V,E)$ be an infinite ordinary tree. Since Ω is an open region, the procedure above to build Kaplan diagrams for finite trees can be repeated countably many times. The Kaplan diagram so obtained has the associated tree similar to T.

Notice that the Kaplan diagram obtained from a tree is by no means unique. Q.E.D.

3.4 Chordal systems and $\mathbb{R}$-trees

We consider arbitrary foliations of the plane in this section.

In order to study the most general situation, we need to recall the concepts of abstract chordal systems and normal chordal systems defined by Kaplan [K1]. See also 1.4.10.

Definition 3.4.1 (2.3, [K1]). Let C be a nonempty set. Assume that there is a set of two relations $a|b|c$ and $|a,b,c|^+$ defined for the different triples of distinct elements a, b, c in E.

C is called a *chordal system*, or a CS if for any triple of distinct elements a, b, c in C exactly one of the following five relations holds: $a|b|c$, $b|c|a$, $c|a|b$, $|a,b,c|^+$, $|a,c,b|^+$ such that the following axioms hold:

(i) $a|b|c$ is equivalent to $c|b|a$.

(ii) $|a,b,c|^+$ is equivalent to $|b,c,a|^+$ (and hence to $|c,a,b|^+$).

(iii) $|a,b,c|^+$ and $|a,c,d|^+$ imply $|a,b,d|^+$ and $|b,c,d|^+$.

(iv) $|a,b,c|^\pm$ and $a|b|d$ imply $c|b|d$ and $|a,d,c|^\pm$, whereby $[a,b,c] \sim [a,d,c]$.

(v) $a|b|c$ and $b|c|d$ imply $a|b|d$ and $a|c|d$.

(vi) Of the three relations $b|a|c$, $b|a|d$ and $c|a|d$ at most two hold.

Here we have the following notations: $|a,b,c|^-$ means $|a,c,b|^+$. $|a,b,c|^\pm$ means $|a,b,c|^+$ or $|a,b,c|^-$. $[a,b,c] \sim [a',b',c']$ means the relation for the triplet a,b,c is the same for a',b',c'.

In this section C will always stand for a chordal system.

A subset U of C is *cyclic* if it contains at least one element and, when U contains at least three elements, for every triple a,b,c in U either $|a,b,c|^+$ or $|a,b,c|^-$ holds.

Lemma 3.4.2. *Let* (T,d) *be a nondegenerate* $\mathbb{R}$*-tree. Let* x_1, x_2 *and* x_3 *be three distinct points in* T. *If* x_1, x_2, *and* x_3 *are not contained in a segment in* T, *then there is a unique* $x \in B_T$ *such that the segments connecting* x *to* x_1, x *to* x_2 *and* x *to* x_3 *intersects only at* x.

Proof. Let $\mu_i\colon [0,\ell_i] \to T$, $i=0,1$, be two closed parameterized segments connecting x_1 to x_2 and x_1 to x_3, respectively. Let $\ell_0 = \sup\{t \mid \mu_1(t) = \mu_2(t)\}$. Then $0 < \ell_0 < \min\{\ell_1, \ell_2\}$. In fact any one of $\ell_0 = 0$, ℓ_1 or ℓ_2 would imply that $\bar{\mu}_1 \cup \bar{\mu}_2$ is a segment containing x_1, x_2 and x_3 (Definition 2.2.4(ii),(iii)). Put $x = \mu_1(\ell_0)$. The rest of the assertion in the lemma is obvious. See the end of the proof of Lemma 2.3.4. Q.E.D.

3.4.3 Let (T,d) be a nondegenerate $\mathbb{R}$-tree. Assign an arbitrary cyclic order on E_x for each $x \in B_T$. Now we define two relations on every triplet of points x_1, x_2, x_3 in T as follows. If there is a segment containing x_1, x_2 and x_3 such that x_2 separates x_1 and x_3, we define $x_1|x_2|x_3$. If x_1, x_2 and x_3 are not contained in any segment in T, by Lemma 3.4.2, there is a unique $x \in B_T$ such that the segments connecting x to x_1, x to x_2 and x to x_3 represent three distinct geodesic directions, say $y_1, y_2, y_3 \in E_x$. We define the relation of the triplet x_1,x_2,x_3 to be the cyclic relation of y_1, y_2 and y_3 in E_x.

Theorem 3.4.4. *With the relations defined as in 3.4.3, the $\mathbb{R}$-tree T becomes an abstract chordal system.*

Proof. The axioms (i), (ii), (v) and (vi) in Definition 3.4.1 hold obviously. We are left with (iii) and (iv).

Suppose that we have $|x_1,x_2,x_3|^+$ and $|x_1,x_3,x_4|^+$. For any two points $x, x' \in T$, we shall denote by $I_{x,x'}$ $(\mathring{I}_{x,x'})$ the closed (open) segment connecting x and x'. From 3.4.3, there are two points $x, x' \in B_T$ and edges $y_1, y_2, y_3 \in E_x$, $y_1', y_3', y_4' \in E_x$ such that $|y_1,y_2,y_3|^+$ and $|y_1',y_3',y_4'|^+$. Here I_{x,x_i} represents y_i, $i = 1,2,3$ and $I_{x',x}$ represents y_i', $i = 1,3,4$. Since both x and x' are in $\mathring{I}_{x_1,x_3}$, there are three possibilities:

(i) $x = x'$. Thus $y_i = y_i'$ for all i and $|y_1,y_2,y_4|^+$, $|y_2,y_3,y_4|^+$ hold. So we have $|x_1,x_2,x_4|^+$ and $|x_2,x_3,x_4|^+$.

(ii) $x' \in \mathring{I}_{x,x_3}$ (see Fig. 3.21). We have $[x_1,x_2,x_4] = [y_1,y_2,y_3]$, $[x_2,x_3,x_4] = [x_1,x_3,x_4]$. Therefore $|x_1,x_2,x_4|^+$ and $|x_2,x_3,x_4|^+$.

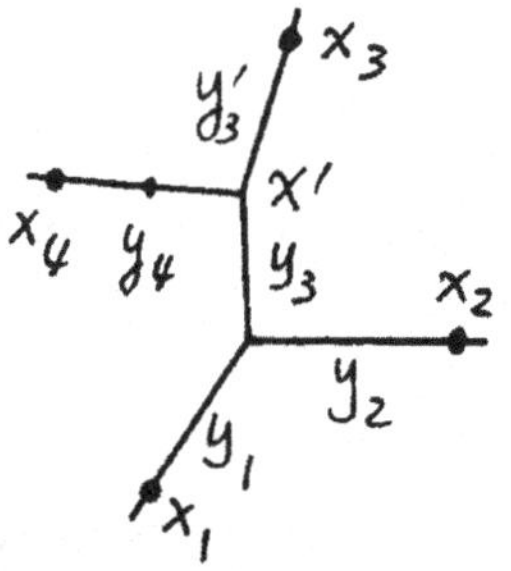

Fig. 3.21

(iii) $x' \in \mathring{I}_{x,x_1}$. Similar to (ii). Thus axiom (iii) holds.

Suppose that $|x_1,x_2,x_3|^+$ and $x_1|x_2|x_4$. Then it is easy to check that $x_3|x_2|x_4$ and $|x_1,x_4,x_3|^+$. See Fig. 3.22. We omit the details. Thus axiom (iv) holds. Q.E.D.

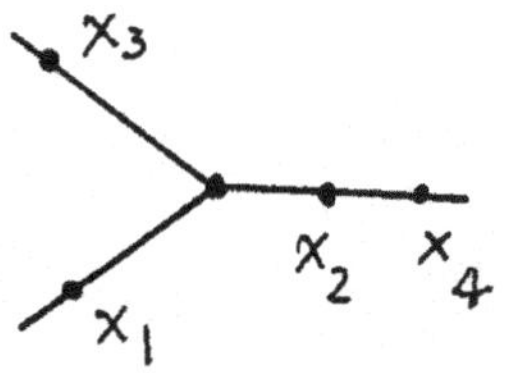

Fig. 3.22

Proposition 3.4.5. *Let* C *be any abstract chordal system and* C_0 *any nonempty subset of* C. *Then* C_0 *is also an abstract chordal system with the relations inherited from* C.

Proof. Check axioms (i)-(vi) in Definition 3.3.1. They are all obvious. Q.E.D.

Naturally C_0 is called a subchordal system of C.

Corollary 3.4.6. *Suppose* T_0 *is a subset of any* $\mathbb{R}$*-tree* (T,d). *Then with the chordal relations defined in 3.4.3,* T_0 *is an abstract chordal system.*

Proof. This follows from Theorem 3.4.4 and Proposition 3.4.5. Q.E.D.

3.4.7 Two chordal systems C_1 and C_2 are *isomorphic* if there is a bijective map from C_1 to C_2 preserving the chordal relations.

A chordal system C is *parallel* if it is isomorphic to the chordal system of (Ω,triv). It is *half-parallel* if it is isomorphic to the chordal system of (Ω_+,triv), where $\Omega_+ = \Omega \cap \{y \geq 0\}$.

An element a of C is said to *divide* C if there exists two subsets $\delta(a)$ and $\delta'(a)$ of C such that $C = \delta(a) \cup \delta'(a) \cup \{a\}$ as a disjoint union and further for b in $\delta(a)$, c in $\delta'(a)$, $b|a|c$; conversely if $b|a|c$ then either $b \in \delta(a)$ and $c \in \delta'(a)$ or else $b \in \delta'(a)$, $c \in \delta(a)$.

A crucial property of chordal systems is that every element of C divides C in a unique way (Theorem 26, [K1]).

Let U be a cyclic set containing at least two elements. Then $\delta(a)$ can be chosen for each a in U so that $U-a \subset \delta(a)$ (Thm. 27, [K1]).

Denote $\theta(U) = \bigcap_{a\in U} \delta(a)$ and call $\theta(U)$ the *interior of the region bounded by* U. Denote $\lambda(U) = \theta(U) \cup U$. Two sets $\lambda(U_1)$ and $\lambda(U_2)$ are *adjacent* if $\lambda(U_1) \cap \lambda(U_2)$ has exactly one element (which divides $\lambda(U_1) \cup \lambda(U_2)$).

Definition 3.4.8. Let A be a countable family of finite sequences α of positive integers. If $\alpha = (n_1, n_2, \ldots, n_\ell) \in A$, then (α,k) stands for the sequence $(n_1, \ldots, n_\ell, k)$. The family A is *admissible* if the following holds:

(1) The sequence $(1) \in A$ and no other sequence is of length 1.

(2) If $(\alpha,k) \in A$, $k > 1$, then $(\alpha,k-1) \in A$.

(3) If $(\alpha,1) \in A$, then $\alpha \in A$.

If A is admissible, there is an ordinary tree associated to A with vertices $\{C_\alpha\}_{\alpha\in A}$ in one-to-one correspondence with A and edges $\{[C_\alpha, C_{(\alpha,k)}], [C_{(\alpha,k)}, C_\alpha]\}_{(\alpha,k)\in A}$. We denote this tree by KT(A) and call it the *Kaplan tree of* A ([K1], p.173).

Definition 3.4.9. A subset E_0 of a chordal system is *seminormal* if there exists an admissible family $A = \{\alpha\}$ and a subset C_0 of C indexed by A, $C_0 = \{a_\alpha\}_{\alpha\in A}$ such that

(1) $C_0 = \delta(a_1) \cup a_1$ for appropriate choice of $\delta(a_1)$;

(2) the sets $U_\alpha = a_\alpha \cup \bigcup_k a_{(\alpha,k)}$ are cyclic;

(3) if α and (α,k) are in A, then $\lambda(U_\alpha)$ and $\lambda(U_{\alpha,k})$ are adjacent;

(4) $a_\alpha \cup \theta(U_\alpha)$ is half-parallel;

(5) $C_0 = \bigcup_\alpha \lambda(U_\alpha)$;

(6) for each element a of $U_\alpha - a_\alpha$ there is an element b of $\theta(U_\alpha)$ such that $|a_\alpha, a, b|^{\pm}$.

The sets $\{U_\alpha\}_{\alpha\in A}$ are then said to determine a *seminormal subdivision* of $C_0^{\bullet}$.

A chordal system C is *normal* if there is an element a of C such that $\delta(a)\cup a$ and $\delta'(a)\cup a$ are seminormal. If these two sets are seminormally divided, then C is *normally* subdivided.

A central result of Kaplan's theory can be stated as follows (Theorem, p. 11 [K2]).

Theorem 3.4.10. *Let* CS(F) *be the abstract chordal system for a foliation* (Ω,F) *defined as in 1.4.10 and 1.2.2. Then* CS(F) *is normal. Conversely let* C *be a normal chordal system. Then there exists a foliation* (Ω,F) *whose chordal system is isomorphic to* C. *Two foliations* (Ω,F) *and* (Ω,F') *are conjugate iff the two chordal systems* CS(F) *and* CS(F') *are isomorphic.*

Example 3.4.11. To explain where the local triviality of a foliation (Ω,F) shows up in the definition of normal chordal systems, let us consider the example shown in Fig. 1.12(a). Let a_α be the chord containing 0. Then for any cyclic set $U_\alpha = a_\alpha \cup \bigcup_k a_{(\alpha,k)}$, where $a_{(\alpha,k)}$ are some chords in the upper disc, $a_\alpha \cup \theta(U_\alpha)$ can never be half-parallel (Axiom (4) of seminormalness).

3.5 Distinguished trees are associated with foliations

Throughout this section unless otherwise stated $T = (T,d,V,E)$ will stand for a distinguished tree. We aim to show that there exists a foliation (Ω,F) with T_2-separatrices such that its associated tree $T(F)$ is isomorphic to T. Our strategy is to build a normal chordal system C for T and then apply Kaplan's result (Theorem 3.4.8) to get $C = CS(F)$ for some foliation (Ω, F). Finally we show that the associated tree $T(F)$ is isomorphic to T.

3.5.1 Recall from Theorem 3.4.5 that any $\mathbb{R}$-tree (T,d) becomes a chordal system if we have assigned a cyclic order on each E_x for each $x \in B_T$.

For a distinguished tree T we shall build a different chordal system $C(T)$ which in a sense is "dual" to the one above. We now make this clearer. Recall (3.4.6) that a parallel chordal system is a chordal system isomorphic to that of (Ω,triv). If we consider the open unit interval $(0,1)$ to be an $\mathbb{R}$-tree, then it is a parallel chordal system (Theorem 3.4.4). If $T = (T,d,V,E)$ is a simplicial tree, then the interior of every (unoriented) geometrical edge is a parallel chordal system with respect to the chordal relations defined in 3.4.4. However if both V and E are $\mathbf{Z}_2$-graded so T becomes a distinguished tree, we can associate another chordal system $C(T)$ to T. To every $x \in V$ a parallel chordal system C_x is attached (instead of to an edge as above), while to every edge $y \in E$, a single chord is associated to it (instead of to a vertex above).

The motivation for the following construction of $C(T)$ is easy to see. It is just the reverse of the construction of a distinguished tree from a foliation (Ω, F), or equivalently from a Kaplan diagram.

Let $T = (T,d,V,E)$ be an arbitrary distinguished tree. As an abstract set, the chordal system $C(T)$ is the quotient set of the collection which consists of the following elements with a certain identifying relation $\sim$. For each $x \in V$ there is a parallel chordal system $C_x \subset C(T)$. For each $y \in E$, with $\deg y = 0$, and each $z \in V'$, there are chords C_y and C_z in $C(T)$. The relation $\sim$ is given as follows.

(1) Recall Lemma 2.3.4. Let $y \in E_x$, with $\deg y = 0$, and $\gamma(y) = (x, x_y)$. If $x_y \in V$ then there is an inverse $\bar{y}$ of y. We let $C_{\bar{y}} \sim C_y$. If $x_y \in V'$, then let $C_y \sim Cx_y$.

(2) If $x, x' \in V_0 \cap B_T$, $\gamma(y) = (x, x')$ and $\deg y = \deg \bar{y} = 1$ (see Fig. 3.2(iv)), then C_x is identified with $C_{x'}$. Now as a set

$$C(T) = \left(\bigcup_{x \in V} C_x\right) \cup \left(\bigcup_{\substack{y \in E \\ \deg y = 0}} \{C_y\}\right) \cup \left(\bigcup_{z \in V'} \{C_z\}\right) / \sim$$

3.5.2 Attaching a half open segment $[0,1)$ to every $x \in V_1$ by $0 \sim x$, we enlarge the distinguished tree T to a new regular $\mathbb{R}$-tree $\tilde{T}$ which will be called the *augmented tree of* T. The vertices of $\tilde{T}$ are the vertices of T. Recall Theorem 3.4.4. After being assigned an arbitrary cyclic relation on each E_x, for $x \in V_0$, and on $\tilde{E}_x = E_x \cup \{y_{x,\infty}\}$ for each $x \in V_1$, the $\mathbb{R}$-tree $\tilde{T}$ becomes an abstract chordal system. Here the extra edge $y_{x,\infty}$ stands for the extra direction represented by the $[0,1)$ attached.

We define an embedding ϕ from the set $C(T)$ into $\tilde{T}$. Thus $C(T)$ is naturally a chordal system itself (Corollary 3.4.6).

(1) If $z \in V'$, then naturally $\phi(C_z) = z$ (see Fig. 3.23).

(2) If $y \in E_x$, $\deg y = 0$, $x_y \in V$, then $\phi(C_y) = \phi(C_{\bar{y}}) = p_y$ is any fixed point in the open segment $\mathring{I}_{x,x_y}$. (If $x_y \in V'$, then $C_y = Cx_y$, which was given in (1).)

(3) If $x \in V_1$, then ϕ identifies C_x with the open segment $(0,1)$ attached to x.

(4) If $x \in V_{0,2}$, there are exactly two edges y_1, y_2 in E_x and ϕ identifies C_x with the open segment $\mathring{I}_{Py_1,Py_2}$.

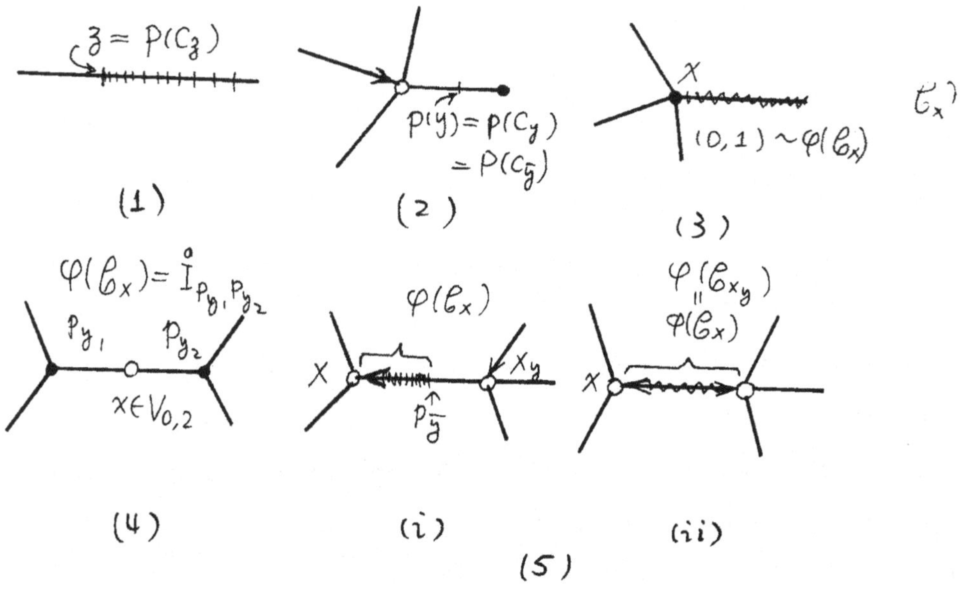

Fig. 3.23

(5) If $x \in V_0 \cap B_T$, there is a unique $y \in E_x$, $\deg y = 1$. There are two possibilities: (i) If $x_y \in V$, $\deg \bar{y} = 0$, then C_x is identified with $\mathring{I}_{x,P_{\bar{y}}}$; (ii) If $x_y \in V$, $\deg \bar{y} = 1$, or $x_y \in V'$, then C_x is identified with $\mathring{I}_{x,x_y}$.

We shall denote by ϕ this embedding of C into $\tilde{T}$.

It is easy to see that ϕ is compatible with the equivalence relation $\sim$ defining $C(T)$. It is also clear that in (3) above a particular choice of the point P_y in $\mathring{I}_{x,x_y}$ does not effect the chordal system $C(T)$. In conclusion we have proved that

Lemma 3.5.3. *For any distinguished tree* T, *the set* $C(T)$ *with the chordal relation defined in 3.5.2, is an abstract chordal system.*

Remark 3.5.4. Clearly the chordal relations on $C(T)$ were determined by that on $\tilde{T}$. It is not hard to show that the converse is also true, although we do not need this. It will become apparent later in this chapter that with cyclic relations assigned at all the vertices of $\tilde{T}$, the distinguished trees determine the conjugacy classes of foliations with T_2-separatrices.

Theorem 3.5.5. *For any distinguished tree* T, *the associated chordal system* $C = C(T)$ *defined in 3.5.2 and 3.5.3 is normal.*

Proof. We shall construct an admissible family A, equivalently, a Kaplan tree, for a seminormal subdivision of the chordal system C (Definition 3.4.8). For the motivation of the construction, see the proof of Theorem 38 [K1].

Fix $C_1 = C_{y_0} \in C$, for some $y_0 \in E$. We have a division $C = D(C_1) \cup D'(C_1) \cup \{C_1\}$ (see 3.4.7). Let $0 \in \tilde{T}$ be the point associated to C_1 (see 3.5.2). Then the point 0 also divides the augmented tree $\tilde{T}$ with $\tilde{T} = \delta(0) \cup \delta'(0) \cup \{0\}$ in such a way that if $C \in D(C_1)$ then the point associated to C is in $\delta(0)$. (The distinguished tree T is regarded as contained in the $\mathbb{R}$-tree $\tilde{T}$.) We choose D and δ so that for any other $C \in C$ and $x \in T$, $D(C_1) \cap D(C) \neq \emptyset$ and $\delta(0) \cap \delta(x) \neq \emptyset$.

Let μ be a parameterized segment with range in $\delta(0) \cup \{0\} \subset \tilde{T}$, with domain of form $(0,\ell)$, where $\ell \in (0,\infty]$ and $\mu(0) = 0$, satisfying the following conditions:

(1) $\bar{\mu} \cap V_1$ contains at most one vertex (where we denote by $\bar{\mu}$ the range of μ and by $\mathring{\mu}$ the interior of $\bar{\mu}$). If $\{x\} = \bar{\mu} \cap V_1$, then $\phi(C_x) \subset \bar{\mu}$. See Fig. 3.24.

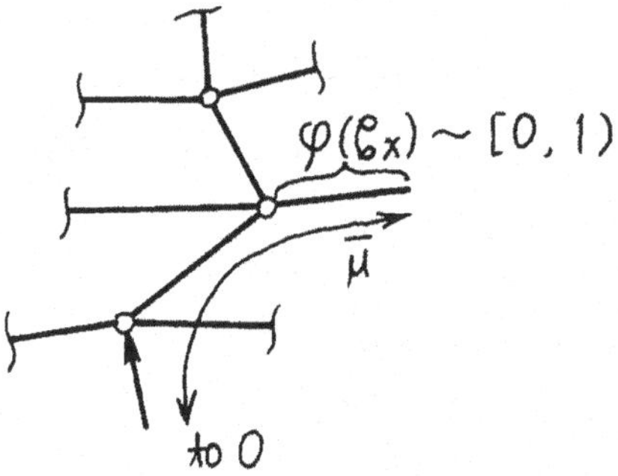

Fig. 3.24

(2) If $x_0 \in \mathring{\mu} \cap B_T \cap V_0$ then $\bar{\mu}$ contains the arrow pointing to x_0 in the sense of Proposition 3.2.18.

(3) If $\mathring{\mu} \cap \Gamma_z \neq \emptyset$, for some $z \in V'$ (see Remark 3.3.2) and $x \in \mathring{\mu} \cap \Gamma_z$, then $\delta(x) \cap \Gamma_z \subset \bar{\mu}$.

(4) μ is not a restriction of a proper extension μ' with domain within $[0,\infty)$ satisfying (1), (2) and (3).

It is easy to see that such a parameterized segment μ exists. In fact we may consider the collection of parameterized segments satisfying (1), (2) and (3) and pick out by Zorn's lemma a maximal element under the partial ordering of inclusion. Moreover $\bar{\mu} \cap V_1 = \emptyset$ iff $\bar{\mu}$ is contained in T (cf. Fig. 1.10). We shall call $\bar{\mu}$ an *extended segment starting at* 0 (or crossing C_1) (compare with Fig. 3.26). Obviously, there is also such a parameterized segment μ' starting at 0 but contained in $\delta'(0) \cup \{0\}$. The union $\bar{\mu}' \cup \bar{\mu}$ shall be called an open extended segment containing 0.

Let $C(\bar{\mu}) = \{c \in D(C_1) \mid \phi(c) \in \bar{\mu}\}$. Recalling (3.5.2) the embedding ϕ and how the chordal relations were defined on C, we see that $C(\bar{\mu})$ is half-parallel (3.4.5). In fact $J = \phi(C(\bar{\mu}))$ is such a subset of the arc $\bar{\mu}$ that: (1) the starting point 0 of $\bar{\mu}$ is in J; (2) for any $x \in J$, $x \neq 0$, either there is a neighborhood of x in $\bar{\mu}$ contained in J or there is an open arc containing x of form given in Fig. 3.25 where $I_{x_1,x_2} \cap J = \{x\}$, and both x_1, x_2 are extremities of two open intervals contained in J. Thus as an ordered set J is isomorphic to $[0,\infty)$.

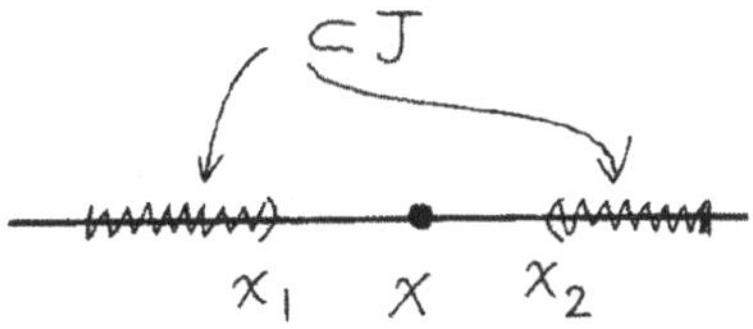

Fig. 3.25

Define a set of chords

$$U_{(1)} = \{C_1\} \cup \{c \in C \setminus C(\bar{\mu}) \mid \phi(c) \in E_x,\ x \in V \cap \bar{\mu}\}.$$

Geometrically if $T = T(F)$ on (Ω, F), these chords correspond to the separatrices (limit separatrices or not) bounding the "semi-maximal region of parallel foliation" whose image under the Kaplan transform Δ is just $C(\bar{\mu})$. The arc $\bar{\mu}$, or rather J, as representative of the chords of $C(\bar{\mu})$, can be regarded as a transversal (Fig. 3.26). Note that for any triple $c_1, c_2, c_3 \in U_{(1)}$, the three

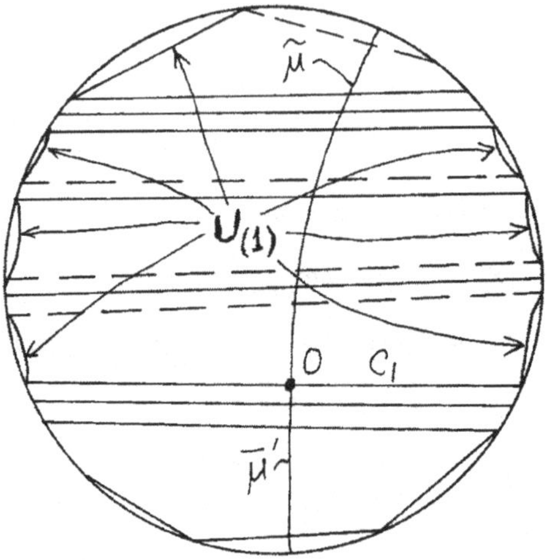

Fig. 3.26

points $\phi(c_1)$, $\phi(c_2)$ and $\phi(c_3)$ in $\tilde{T}$ are not contained in any single segment (3.5.2). Thus $U_{(1)}$ is cyclic (3.4.1, 3.4.3). Moreover the set of chords $\theta(U_{(1)}) = D(c_1) \cap C(\bar{\mu})$ is parallel.

Recalling the beginning of this proof, we now construct an admissible family A. We define $A_1 = \{1\}$. Enumerate $U_{(1)}$ by $c_1, c_{(1,1)}, c_{(1,2)}, \ldots, c_{(1,k)}$, $k = 1, \ldots, \infty$. We define $A_2 = \{(1,1),(1,2),\ldots,(1,k)\}$. By induction, suppose that at the nth step we have defined a collection of chords $\mathcal{U}_n = \{c_\alpha\}_{\alpha \in A_n}$, where each sequence α is of length n. At the (n+1)-th step, we fix each $c_\alpha \in \mathcal{U}_n$ and replace c_1 by c_α in the procedure above and obtain a cyclic set $U_\alpha = \{c_\alpha, c_{(\alpha,1)}, c_{(\alpha,2)}, \ldots\}$. We then define

$A_{n+1} = \{(\alpha,k) | c_{(\alpha,k)} \in U_\alpha, \alpha \in A_n\}$ and $\mathcal{U}_{n+1} = \{c_\alpha\}_{\alpha \in A_{n+1}}$. To show that $C(T) = D(c_1) \cup \{c_1\} \cup D'(c_1)$ is a normal division, it suffices to show that the family $A = \bigcup_{n=1}^{\infty} A_n$ of indices is admissible (Definition 3.4.8 and Definition 3.4.9).

Let $\lambda(U_\alpha) = \theta(U_\alpha) \cup U_\alpha$ for all $\alpha \in A$ (see 3.4.7). Then clearly $\lambda(U_\alpha) \cap \lambda(U_{(\alpha,k)}) = c_{(\alpha,k)}$, $(\alpha,k) \in A$. So $\lambda(U_\alpha)$ and $\lambda(U_{(\alpha,k)})$ are adjacent.

All the axioms (1)-(4) of Definition 3.4.9 are clear at this point. Axiom (6) is also obvious from 3.5.2. We need to show

$$D(c_1) \cup \{c_1\} = \bigcup_{\alpha \in A} \lambda(U_\alpha) . \tag{3.11}$$

Denote the set of chords on the right hand side of (3.11) by C_0. Let $\tilde{T}(0) = \delta(0) \cup \{0\} \subset \tilde{T}$. Let $\tilde{T}_0$ be all the points x in $\tilde{T}(0)$ such that

(i) $x = \phi(c)$ for some $c \in C_0$;

(ii) there is an arc I in $\tilde{T}_0$ containing x and some element satisfying (i), with $\mathring{I} \cap \bar{V} = \emptyset$.

Looking closely at the construction in 3.5.2, we note that $B_T \cap \phi(C(T)) = \emptyset$. A moment's reflection on (i) and (ii) and the definition of $\tilde{T}_0$ tells us that a point x in $\tilde{T}(0)$ is contained in $\tilde{T}_0$ iff there is an extended segment μ_α starting at $\phi(c_\alpha)$, $\alpha \in A$, such that either $x \in \bar{\mu}_\alpha$, or x is in the open segment connecting some $x_0 \in B_T \cap \bar{\mu}_\alpha$ and $P(c_{(\alpha,k)})$, $(\alpha,k) \in A$, and this open segment contains no point in $\bar{V}$ (see Fig. 3.27).

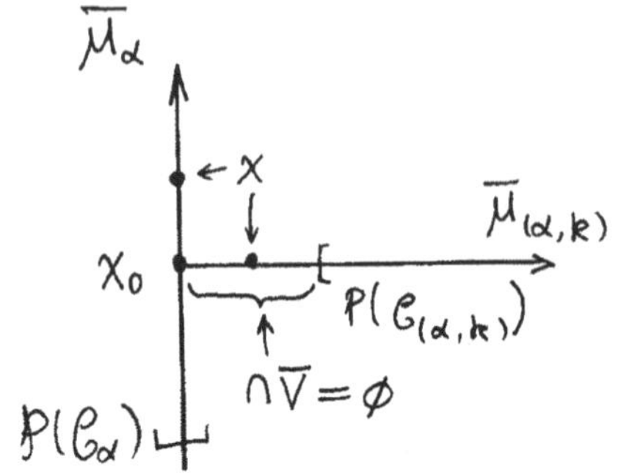

Fig. 3.27

We claim that $\tilde{T}_0$ is a closed subset of $\tilde{T}(0)$. So let $x_0 \in \bar{\tilde{T}}_0 \setminus \tilde{T}_0$. Note first that x_0 is in $\bar{V}$. Otherwise x_0 is in the open set $\tilde{T}(0)\setminus(\bar{V}\cup\{0\})$. So there is a segment I_0 containing x_0 such that $\mathring{I}_0 \cap \bar{V} = \emptyset$, and $I_0 \cap \tilde{T}_0$ contains some element x. If x satisfies (i), then x_0 is in $\tilde{T}_0$ by (ii). Let x satisfy (ii), i.e., there is an arc $I \subset \tilde{T}(0)$ containing x and some element x' satisfying (i), with $\mathring{I} \cap \bar{V} = \emptyset$. Let $I' = I \cup I_0$. Then I' is an arc in $\tilde{T}(0)$ containing both x_0 and y, with $\mathring{I}' \cap \bar{V} = \emptyset$. Again x_0 satisfies (ii) and $x_0 \in \tilde{T}_0$.

Note next that if $x_0 \in V \cap \bar{\tilde{T}}_0$ then there is some $\mathring{I}_{x,x_y}$ containing a point $x \in \tilde{T}_0$. Replacing I_0 by I_{x,x_y} above, we see that $x_0 \in \tilde{T}_0$.

We are left with the case $x_0 = z \in V'$. There is an arc Γ_z containing z in its interior (Remark 3.3.2) satisfying axiom (v), Definition 3.3.1. From the discussion above we see that if a point on a geometric edge is in $\tilde{T}(0)$ (a geometric edge means an arc of form I_{x,x_y}, $y \in E_x$) then both the extremities of this geometric edge are in $\tilde{T}(0)$. So there is a sequence $\{x_n\} \subset V_0 \cap B_T \cap \tilde{T}$ converging to x_0 from one side. If $\{x_n\} \subset \delta'(z)$, then an extended segment μ passing x_n, for n large, will also pass z. Thus z will satisfy (i) and $z \in \tilde{T}_0$. Now we assume $\{x_n\} \subset \delta(z)$ (see Fig. 3.28). We may assume $\{x_n\} \subset \Gamma_z \cap B(z, \ell_z)$. By our remark earlier

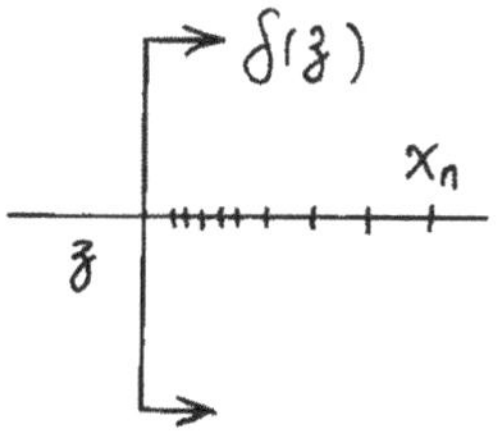

Fig. 3.28

about $\tilde{T}_0$ (Fig. 3.27), each of these x_n is in an extended segment $\bar{\mu}_{\alpha_n}$, $\alpha_n \in A$. By condition (3) of the definition of an extended segment, all these $\bar{\mu}_{\alpha_n}$ coincide for various n, as they all intersect with Γ_z. So we may write $\bar{\mu} = \bar{\mu}_{\alpha_n}$. Thus z is in the closure of $\bar{\mu}$ and hence contained in $\bar{\mu}$ as $\bar{\mu}$ is closed in $\bar{T}(0)$. Thus $\tilde{T}_0$ is closed in $\tilde{T}(0)$.

One verifies also that $\tilde{T}_0$ is open in $\tilde{T}(0)$. Consider again Fig. 3.27. Recall the remark about $\tilde{T}_0$ there. If x is contained in a geometric edge of form $I_{x,x_y} \subset \tilde{T}_0$, for I_{x,x_y} in an extended segment or not, then x is of course an interior point of T_0. If $x \in \tilde{T}_0 \cap V$, then x has to be an interior point of some extended segment $\bar{\mu}$, and every geometrical edge I_{x,x_y}, $y \in E_x$, except the two lying in $\bar{\mu}$, contains the starting point of a new extended segment. Thus x is again an interior point of $\tilde{T}_0$. Finally let $x \in V' \cap \tilde{T}_0$. By the above discussion, there is an extended segment $\bar{\mu} \subset \tilde{T}_0$ containing $B(z,\ell_z) \cap V$. Thus z is also an interior point of $\tilde{T}_0$.

Therefore $\tilde{T}_0$ is both open and closed in $\tilde{T}(0)$, and is thus $\tilde{T}(0)$ itself, as $\tilde{T}(0)$ is connected. We have $\phi(C_0) \subset \tilde{T}(0)$ and $\phi(D(c_1) \cup \{c_1\}) \subset \tilde{T}(0)$. Recall again the remark earlier about the structure of $\tilde{T}_0$ ($= \tilde{T}(0)$); see Figs. 3.27 and 3.23. If $x = \phi(c)$, for some $c \in D(c_1) \cup c_1$, then x is necessarily contained in some extended segment $\bar{\mu}_\alpha$, $\alpha \in A$, and thus $c \in \lambda(U_\alpha)$. So in the inclusion $\phi(C_0) \subseteq \phi(D(c_1) \cup \{c_1\})$, the equality holds. Thus (3.11) holds as ϕ is injective. It follows that $D(c_1) \cup \{c_1\}$ is seminormal and so $D'(c_1) \cup \{c_1\}$ is. The chordal system $\mathcal{C}$ is normal. Q.E.D.

3.5.6 The proof of Theorem 3.5.4 exhibited the relation between a distinguished tree T and the Kaplan tree $KT(\Sigma)$ arising from a normal subdivision Σ (not unique) of a chordal system (also not unique) $\mathcal{C}$ associated to T. Roughly the construction of $KT(\Sigma)$ is as follows. Suppose $\mathcal{C} = \mathcal{C}(T)$ has been chosen for T. Note that the isomorphism class of $\mathcal{C}$ depends only on: (1) the similarity class of T; (2) the cyclic relations on each E_x, $x \in V_0$, and $\tilde{E}_x$, $x \in V_1$ (3.5.2). Fix an arbitrary geometrical edge $y \in E$ (the nonuniqueness of a normal subdivision starts coming in here). There is a point $P_y = 0$ associated to C_y. Recall the construction of an index set $A = \{\alpha\}$ of finite sequences of positive integers, for "half" of the chordal system, $D(C_y) \cup \{C_y\}$, or equivalently, for half of the distinguished tree, $\delta(0) \cup \{0\} = T(0)$. We have correspondingly "half" of the augmented tree $\tilde{T}$, i.e. $\tilde{T}(0)$. Let S be the union of all the extended segments $\bar{\mu}_\alpha$, $\alpha \in A$. Then $\tilde{T}(0)/S$ is a quotient $\mathbb{R}$-tree (2.4.3). Define the set of vertices of $\tilde{T}(0)/S$ to be the set of extended segments. Then the geometrical edges of $\tilde{T}(0)/S$ (which are represented by the points $P(C_\alpha)$, $\alpha \in A$ on $\tilde{T}(0)$) are in one-to-one correspondence with the vertices of $KT(A)$ (Definition 3.4.8), while the edges of $KT(A)$ can be mapped (not one-to-one!) onto the vertices of $T(0)/S$ in the following way: every edge of form $(P(c_\alpha), P(c_{(\alpha,k)}))$ or $(P(c_{(\alpha,k)}), P(c_\alpha))$ is mapped to $\bar{\mu}_\alpha$, $\alpha \in A$. Let $KT(A')$ be the Kaplan tree for an admissible family $A' = \{\alpha'\}$ associated to the chordal system $D'(C_y) \cup \{C_y\}$. Then we have a Kaplan tree $KT(\Sigma) = KT(A) \cup KT(A')/\sim$ for the distinguished tree T.

Although we mentioned the chordal system while constructing the Kaplan tree $KT(\Sigma)$, from the discussion above, clearly we see that $KT(\Sigma)$ depends only on the choice of extended segments $\{\bar{\mu}_\alpha\}$, $\alpha \in A$, which is intrinsic to T. In fact the definition of extended segment depends on a subset $\phi(C)$ of T (3.5.2), which is intrinsic to T, i.e., independent of the cyclic relations which have been assigned at the vertices of $\tilde{T}$.

Example 3.5.7. In Fig. 3.29 we show that two different choices of extended segments of the same distinguished tree yield nonisomorphic Kaplan trees.

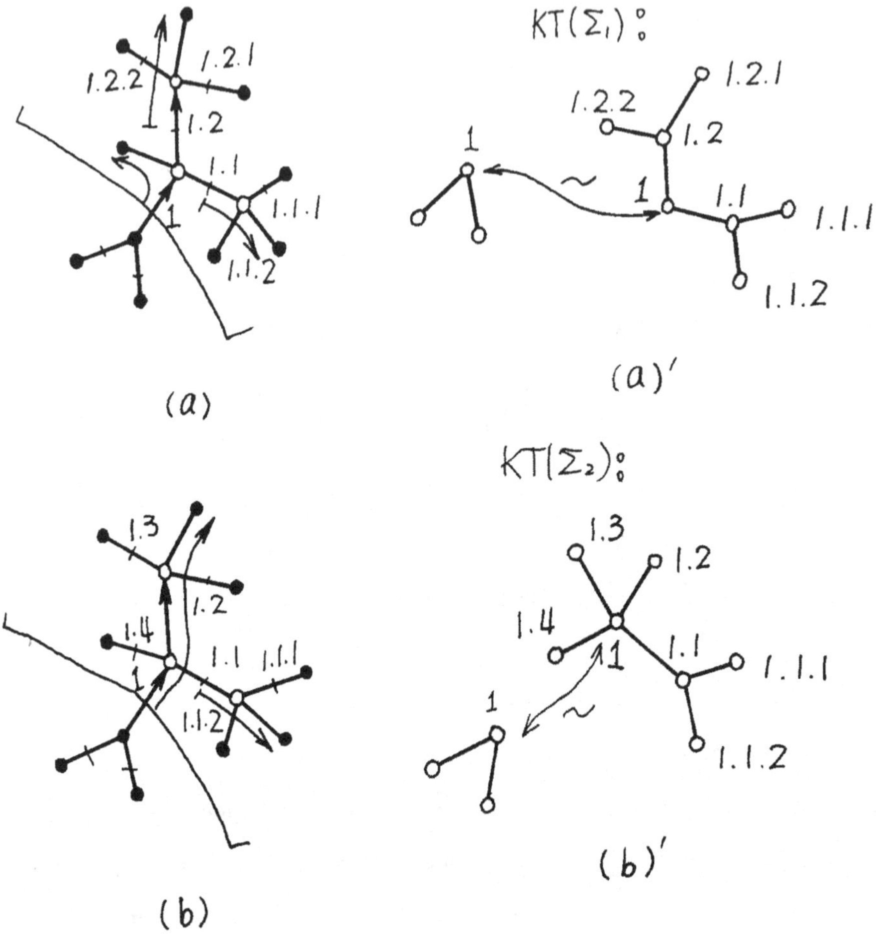

Fig. 3.29

Recall that every extended segment corresponds to a collection of edges meeting at the vertex representing the starting point of the extended segments. Some extended segments are shown in (a) and (b). The extended segments can be represented by their starting points $p(c_\alpha)$ (3.5.5) which are in turn represented by the vertices in the Kaplan trees, as shown in Fig. 3.29.

The results we obtained in 3.5.2-3.5.5 suggest an equivalent definition of distinguished trees.

Definition 3.5.8. A *quasi-distinguished tree* is a regular $\mathbb{R}$-tree satisfying axioms (i)-(iv) in Definition 3.3.1.

Definition 3.5.9. A *generalized extended segment starting at* $x \in T$ is a parameterized segment μ with domain of form $[0,\ell]$ or $[0,\ell)$, $\ell \in (0,\infty]$, $\mu(0) = x$, satisfying the following conditions:

(1) $\mathring{\mu} \cap V_1 = \emptyset$

(2) If $x_0 \in \bar{\mu} \cap B_T \cap V_0$, then $\bar{\mu}$ contains the arrow pointing to x_0 (Proposition 3.2.18).

(3) μ is not a restriction of a proper extension μ' with domain within $[0,\infty)$ satisfying (1) and (2).

Sometimes we also refer to $\bar{\mu}$ as a generalized extended segment if there is no danger of confusion.

Definition 3.5.10. For a quasi-distinguished tree T, a *Kaplan construction* consists of the following procedures:

(1) Fix a point $0 \in T$. Choose a generalized extended segment $\mu_{1,k}$ such that $\mathring{\mu}_{1,k} \subset T_{1,k}$, for each component $T_{1,k}$ of $T\setminus\{0\}$, $k = 1,2,\ldots$. Put $A_1 = \{1\}$, and $A_2 = \{(1,k)\}$ as these index sets.

(2) Suppose at the n-th step we obtain a collection of generalized extended segments, $\{\mu_\alpha\}_{\alpha\in A_n}$, where the α's are sequences of positive integers of length $(n+1)$. At the $(n+1)$-th step, for each $\alpha \in A_n$, each $x \in B_T \cap \bar{\mu}_\alpha$, and each $y \in E_x$ that $\mathring{I}_{x,x_y} \cap \bar{\mu}_\alpha = \emptyset$, we choose a generalized extended segment $\mu_{(\alpha,k)}$ starting at x and containing I_{x,x_y}, $k = 1,2,\ldots$. Let $A_{n+1} = \{(\alpha,k)\}$ be the index set for $\{\mu_{(\alpha,k)}\}$.

(3) Set $A = \bigcup_{n=1}^{\infty} A_n$.

The Kaplan tree KT(A) (3.4.6) is called an *admissible Kaplan tree for* T *if* $\bigcup_{\alpha\in A} \bar{\mu}_\alpha = T$.

Theorem 3.5.11. *A quasi-distinguished tree* T *is a distinguished tree iff*

(5)' T *has an admissible Kaplan tree.*

More precisely, there *exists* a Kaplan construction which yields an admissible Kaplan tree for T.

Proof. The proof of Theorem 3.5.4 gives a Kaplan construction which yields an admissible Kaplan tree for a distinguished tree T.

Conversely, a moment of reflection shows that axiom (5)' above implies axiom (5) in Definition 3.3.1. In fact every $z \in V'$ is contained in some generalized extended segment $\bar{\mu}_z$. We simply take $\Gamma_z = \bar{\mu}_z$. Then the conditions (ii) and (iii) of axiom (5) are automatically true, while (i) of (5) easily follows from the fact that an admissible Kaplan tree is an *ordinary* tree. Q.E.D.

Note that axiom (5) of Definition 3.3.1 is essentially a "local" condition, while axiom (5)' is a "global" one, and Theorem 3.5.11 shows that it turns out to be the same.

Finally we conclude with one of the main results in this section which has become apparent at this stage.

Theorem 3.5.12. *Let* T = (T,d,V,E) *be a distinguished tree. Then there is a foliation* (Ω,F) *with* T_2*-separatrices such that the associated tree* T(F) *is similar to* T.

<u>**Proof**</u>. Let C be a chordal system associated to T, constructed as in 3.5.1, 3.5.2 and 3.5.3. By Theorem 3.5.4, C is normal. From Kaplan's results (Theorem 12 [K2]) there is a foliation (Ω,F) corresponding to C and unique up to conjugacy. Consult section 1 [K2] for the description of (Ω,F).

Let the Kaplan diagram associated to the foliation (Ω,F) be $K(F) = (C',P)$; see §1. One verifies that the chordal system C' is isomorphic to C (Definition 3.4.7). In fact the leaves of F are in one-to-one correspondence with the chords in C (1.7 [K2]) and with those in C' as well (recall the Kaplan transform Δ in 1.2.4). Furthermore, the chordal relations are preserved under the one-to-one correspondence from C to C'. A careful observation of §3.1 (Fig. 3.2) shows that the five types of canonical regions can be determined solely by the chordal relations (including the boundary chords). Thus from (C',P) we recover exactly the vertices V and edges E of (T,d,V,E). The limit chords are also preserved. Let $T(F)$ be the tree associated to (Ω,F). Thus by the construction of $T(F)$ from $K(F)$ described in §3.2 it is clear that $T(F)$ is similar to T.

Let S be the set of separatrices of (Ω,F). Applying a topological conjugacy if necessary (we may assume that every leaf of (Ω,F) has planar measure zero. The Kaplan transform Δ sends S to a set $\Delta(S)$ of chords in $K(F)$. Each chord in $\Delta(S)$ is either a closed side of a polygonoid or a limit chord (Theorem 1.5.7). By Lemma 3.5.12, we can show easily that the set of intersections of $\Delta(S)$ with $\partial\Omega$ has length zero. Thus the planar area of S is zero by the construction of Δ. Applying the imbedding ϕ defined in 3.5.2, the set $\Delta(S)$ is identified with a closed subset of $T(F)$, which is of course Hausdorff with the metric topology. It is easy to see that with the restricted topology, the set S is homeomorphic to the set $\phi(\Delta(S)) \subset T(F)$. Thus S is Hausdorff. Q.E.D.

Lemma 3.5.13. *Let* D *be a discrete subset of* $[0,1]$. *Then the closure* $\bar{D}$ *has Lebesgue measure zero.*

Proof. For any $\varepsilon > 0$, we can find a finite union F_ε of closed subintervals of $[0,1]$, such that (i) $F_\varepsilon \subset [0,1]\backslash D$ and (ii) $m(F_\varepsilon) > 1-\varepsilon$. Q.E.D.

§4. The C*-algebras of foliations of the plane

Let (Ω,F) be an arbitrary foliation of the plane. Recall that S is sometimes the closed subset of separatrices in the leaf space Ω/F and sometimes the closed subset of Ω consisting of all separatrices, depending on the context. We may assume that (Ω,F) arises from a free $\mathbb{R}$-action (1.1.4).

Proposition 4.0.1. *The C*-algebra $C^*(\Omega,F)$ is always CCR. Moreover $spec(C^*(\Omega,F))$ is T_1 and homeomorphic to the leaf space Ω/F with the quotient topology.*

Proof. From Prop. 1.2.1a, the orbit space Ω/F is T_1. By ([Gt] or [Wil]), the C*-algebra $C^*(\Omega,F)$ is CCR and $spec(C^*(\Omega,F)) \simeq \Omega/F$. Q.E.D.

Whenever the foliation (Ω,F) is nontrivial, there are at least two sepratrices, and the $\mathbb{R}$-action is nonwandering, so the C*-algebra $C^*(\Omega,F)$ is not of continuous trace (Thm. 14. [Gr]). It is interesting to find the way $C^*(\Omega,F)$ expressing as composition series with the subquotients being continuous trace C*-algebras [Dix].

By Prop. 1.1. or Prop. 1.5. of Part I, we have an exact sequence

$$(4.0) \qquad 0 \to C^*(U,F) \to C^*(\Omega,F) \to C^*(S,F) \to 0$$

The ideal $C^*(U\ F)$ is a continuous trace C*-algebra. In fact it is isomorphic to $\bigoplus_{i\in I} C_0(\mathbb{R}) \otimes K$, where I is the index set (countable) for canonical regions. The quotient $C^*(S,F)$ is CCR, its composition series of continuous trace determines that of $C^*(\Omega,F)$. It may happen that the index sets for these composition series are of continuum. Consider Example 1.5.10. The candidates of which the C*-algebras have countable index sets of such composition series are the ordinary foliations (1.5.13.). We shall determine explicitly all the C*-algebras with T_2-separatrices, those, we shall show, are exactly that the $C^*(S,F)$ are of continuous traces (4.0). Our method can be generalized to computing arbitrary ordinary foliations. However, to classify these foliations, the combinatorial structure of the corresponding invariant (similar to distinguished trees) will be more complicated.

Unless otherwise stated, "foliation" will still stand for "foliation with T_2-separatrices". By the approach we take of viewing the C*-algebras of foliations of the plane as noncommutative CW 1-complexes. The classification problem is naturally reduced to three problems: the first is to find out what are the 1-cells; the second is the classification of the structure of the noncommutative CW 1-complexes; and the third is to verify that indeed these CW 1-complexes are precisely the C*-algebras of foliations of the plane. These three problems are dealt with in section 4.1, 4.2, and 4.3, respectively.

In §1 we saw that a foliated plane (Ω,F) is decomposed into canonical regions and separatrices. Along with their boundaries, the canonical regions are classified into five conjugacy types as shown in Cor. 1.5.8. Thus C*-algebras of these foliated regions (noncompact, with boundaries) are the building blocks of $C^*(\Omega,F)$.

To find all possible 1-cells, it boils down to computing the C*-algebras for the parallel foliations of the polygonoids. This will be carried out in 4.1. These C*-algebras are naturally 1-cells with various numbers of "branches" at one end, depending on the number of closed sides of the polygonoid.

The "branching" phenomenon happens quite often in geometry, in particular in the theory of singularities, in low dimensional manifolds, and in dynamical systems. It turns out that in C*-algebras it occurs naturally as well. Since in C*-algebras, nonhausdorffness is common rather than exceptional, noncommutative geometry may have its own advantage in studying such phenomenons.

In order to describe the structure of noncommutative CW-complexes (not necessarily "1-complexes"), in 4.1 we introduce the notion of a regular $\mathbb{R}$-trees of C*-algebras, so that these CW-complexes can be regarded as global fibred products associated in trees. We find in particular that the structure of the CW 1-complexes with which we are concerned are precisely determined by the distinguished trees (which were introduced in §3), and vis versa (Thm. 4.2.21).

In 4.3, we prove that these CW 1-complexes determined by distinguished trees are exactly the C*-algebras of foliation with T_2-separatrices (Thm. 4.3.1). By summing up all the previous results, the classification theorem is given in Thm. 4.3.2.

4.1. **C*-algebras of parallel foliations of polygonoids**

"Straightening out" all the leaves of the parallel foliations of a polygonoid (Fig. 3.6.), we see that a foliation polygonoid can be considered as the horizontally constantly foliated closed upper half plane with a closed set of measure zero deleted from its boundary. So far C*-algebras for foliated manifolds with boundaries are not yet defined. Observing in our case the foliations are induced by free actions of a Lie group, we might define the C*-algebras of foliations to be the C*-algebras of corresponding transformation groups. However in this way we would step back to where we begin with, because it is extremely difficult, if not impossible, to compute the C*-algebras of these transformation groups by the definition involving explicit group actions (cf. §5, Part I). Recall that C*-algebras of foliations if closed manifolds are the reduced C*-algebras of holomomy groupoids and C*-algebras of transformation groups are the reduced C*-algebras of transformation groupoids. The setting we shall adopt here is reduced C*-algebras of locally compact groupoids ([Ren]). From this broad point of view Props. 1.1 and 1.5 in Part I share a unified generalization, which will be shown to be crucial rather than just a formalism.

Let $\mathbf{G}$ be a locally compact groupoid with a fixed left Haar system $\{\lambda^u, u \in G^0\}$ (p. 16., [Ren]). We call a subset $\mathbf{G}'$ of $\mathbf{G}$ a *subgroupoid* if $\mathbf{G}'$ is closed under groupoid operation of $\mathbf{G}$ i.e. $x \in \mathbf{G}'$ implies $x^{-1} \in \mathbf{G}'$ and $x,z \in G'$ and x,y composible imply $xy \in \mathbf{G}'$.

Proposition 4.1.1. *Let* $\mathbf{G}$ *be a locally compact groupoid with left Haar* *system* $\lambda = \{\lambda^u, u \in G^0\}$. *Let* G' *be an open subgroupoid of* G *with induced measures* $\{\lambda^u|G', u \in G'^0\}$, *where* $G'^0 = G' \cap G^0$. *Then* $\lambda' = \{\lambda^u|G'^u, u \in G'^0\}$ *is a left Haar system of* G'. *Fix* λ' *to* G'. *The inclusion* $C_c(G') \subset C_c(G)$ *extends to an isometric * homomorphism of* $C^*(G')$ *into* $C^*(G)$.

Proof. Denote $\lambda'^u = \lambda^u|G'$, for $u \in G'^0$. Then obviously the support of measure λ'^u is $G' \cap G^u$. for any function $f \mapsto C_c(G')$, the mapping $u \mapsto \int_{G'} f d\lambda'^u = \int_G f\, d\lambda^u$ is continuous. For any $x \in G'$ and $f \in C_c(G')$, we check easily the left invariant property: $\int_{G'} f(xy)\ d\lambda'^{d(x)}(y) = = \int_G f(xy)\ d\lambda^{d(x)}(y) = \int_G f(y)\ d\lambda^{r(x)}(y) = \int_{G'} f(y)\ d\lambda'^{t(x)}(y)$. To see the first equality, note $x \in G'$, $xy \in G'$ imply $y = x^{-1} \cdot xy \in G'$. The second equality follows from the left invariant property of $\{\lambda^u, u \in G^0\}$. The last equality follows from that $supp(f) \subset G'$ and $\lambda'^{r(x)} = \lambda r^{(x)}|G'$. Therefore $\{\lambda'^u, u \in G'^0\}$ is a left Haar system for G'. The proof of the rest of the conclusion is completely analogous to that of Prop. 1.5., Part I. Q.E.D.

In Prop. 4.1.1, note that the left Haar system fixed to G' is always assumed to be the one induced from that of the ambient groupoid G. However this is not very essential. By Cor. 2.11, P. 85, [Ren], if $\lambda_1 = \{\lambda_1^u, u \in G'^0\}$ is any other left Haar system associated to G', then the reduced C*-algebra of G' corresponding to λ' is strongly Morita equivalent to that of G' corresponding to λ_1 (cf. [Rie], [B-G-R]), and therefore the two C* algebras are isomorphic ([H-S]).

Note also that Prop. 4.1.1 is a "proper" extension of Props. 1.1. and 1.5., because the open subgroupoid $\mathbb{G}'$ is not necessarily arised from a saturated open submanifold (when $\mathbb{G}$ is the holonomy groupoid of a foliation), or an invariant open subset (when $\mathbb{G}$ is the transformation groupoid). As a matter of fact, we are going to use this feature to treat the foliated manifolds with boundaries which we are concerned with.

Theorem 4.1.2. *Let* F *be a closed subset of the real line with Lebesgue measure zero. Let* X_F *be the closed upper half plane with* F *deleted from the boundary. Let* $(X_F, \mathcal{F})$ *be the foliation of* X_F *with horizontal leaves. Define the* C* *algebra of foliation* $(C_F, \mathcal{F})$ *to be the reduced C*algebra of the subgroupoid* $\mathbb{G}_{X_F}^{X_F}$ *of the holonomy groupoid* $\mathbb{G}(\mathbb{R}^2, \mathcal{F})$ *of the horizontally foliated plane. Then*

1) *when* F *is a finite set* n *points, then* $C^*(X_F, \mathcal{F})$ *is naturally isomorphic to*

$$(4.1)\quad A_{n+1} = \{f \in C_0([0,\infty), M_{n+1}(K)) \mid f(0) = \begin{pmatrix} a_0 & & & \\ & \cdot & & \\ & & \cdot & \\ & & & a_n \end{pmatrix}, a_i \in K\}$$

2) *When* F *is infinite, then* $C^*(X_F, \mathcal{F})$ *is naturally isomorphic to*

$$(4.2)\quad A_\infty = \{f \in C_0([0,\infty), K \times K) \mid f(0) = \begin{pmatrix} a_0 & & \\ & a_1 & \\ & & \ddots \end{pmatrix}, a_i \in K\}$$

<u>Remark 1</u>. This theorem can be proved with different method in the situation that the closed set F is not necessarily of measure zero (Thm. 4.1.3). (In fact, for any closed subset F of $\mathbb{R}$ that $\mathbb{R}\setminus F \neq \phi$, there is a closed subset F' of $\mathbb{R}$ of measure zero, such that $(X_F, \mathcal{F})$ and $(X_{F'}, \mathcal{F})$ are topologically conjugate!) However the proof given here has the advantage of being constructive.

<u>Remark 2</u>. Under the topological conjugacy of X_F with its corresponding canonical region Ω_x, an $\mathbb{R}$-action inducing the foliation on Ω will induce an $\mathbb{R}$-action on Ω_x therefore on X_F. This $\mathbb{R}$-action on X_F defines naturally a left Haar system on the transformation groupoid of $(X_F, \mathbb{R})$, which is not equivalent to the left Haar system fixed to $G_{X_F}^{X_F}$ induced by foliation. However as we remarked earlier, the two Haar systems give arise to isomorphic C* algebras.

<u>Proof of Thm. 4.1.2</u>. Let $(H, \mathcal{F})$ be the foliated closed upper half plane with horizontal leaves. As a subgroupoid of $G(\mathbb{R}^2, \mathcal{F})$, the graph is

$$G(H,\mathcal{F}) \cong [0,\infty) \times \mathbb{R}^2 \tag{4.3}$$

with the multiplication rule

$$(t,x,y)(t,y,z) = (t,x,z) \tag{4.4}$$

where t stands for the transversal coordinate.

By Prop. 1.3., Part I

$$C^*(H,F) \cong C_0[0,\infty) \otimes K(L^2(\mathbb{R})) \tag{4.5}$$

Let the canonical isomorphism in (4.5) be ψ. For any $f \in C_c(\mathbb{G}(H,F)) \subset C^*(H,F)$, we may consider that $f \in C_c([0,\infty) \otimes \mathbb{R}^2)$. For each $t \in [0,\infty)$, $\xi \in L^2(\mathbb{R})$,

$$\psi f_t \xi(x) = \int_{\mathbb{R}} f(t,x,y)\xi(y)\, dy \tag{4.6}$$

Now let us consider the foliation (X_F, F). Write $\mathbb{R} \setminus F = \cup_{k=1}^{n+1} U_k$, where $\{U_k\}$ is a collection of disjoint open intervals, n is either finite or $\aleph_0$ depending on F. Let P_k be the operator on $L^2(\mathbb{R})$ of multiplying by χ_{U_k}.

Then

$$L^2(\mathbb{R}) \cong \oplus_{k=1}^{n+1} P_k L^2(\mathbb{R})$$

Accordingly, (4.5) becomes

$$C^*(H,F) \cong C_0[0,\infty) \otimes M_n(K).$$

Here $M_n(K)$ is understood to be $K \otimes K$ if $n = \aleph_0$. The graph of (X_F, F),

$$\mathbb{G}(X_F,F) = \{(t,x,y) \in [0,\infty) \times \mathbb{R}^2 \mid \text{if } t = 0, \text{ then } y, z \text{ lie in the same } U_i \text{ for some } i\}. \tag{4.7}$$

with the same multiplication rule (4.4).

Clearly $G(X_F,F)$ is an open subgroupoid of $\mathbb{G}(H,F)$. By Prop. 4.1.1., the inclusion $C_c(G(X_F,F)) \subseteq C_c(G(H,F))$ extends to an isometric * isomorphism τ from $C^*(X_F,F)$ into $C^*(H,F)$. Let us find the image of τ.

Let $f \in C_c(G(X_F,F))$. Then $\tau f \in C_c(H,F))$ satisfies

$$\tau f(0,x,y) = 0 \quad \text{if} \quad y \in U_i,\ z \in U\ ,\ i \neq j.$$

For simplicity, let us assume in addition that for some fixed i

$$f(0,x,y) = 0 \quad \text{if} \quad x \notin U_i \quad \text{or} \quad y \notin U_i. \tag{4.8}$$

Of course (4.8) holds for τf as well. By (4.6),

$$(\psi\circ\tau f)_0(\xi)(x) = 0, \quad \text{if} \quad x \notin U_i. \tag{4.9}$$

If $\operatorname{supp}(\xi) \subset U_j$, $j \neq i$, then

$$(\psi\circ\tau f)_0(\xi)(x) = 0 \qquad \text{for all } x \in \mathbb{R}. \tag{4.10}$$

From (4.9) and (4.10), it follows that

$$\operatorname{Supp}(\psi\tau f)_0 \subset L^2(U_i),\ \operatorname{rang}(\psi\tau f)_0 \subset L^2(U_i).$$

Let $i = 0, 1, 2, n$ vary. We see that $(\psi\tau f)_0$ is diagonal in $K(\oplus_{i=0}^{n} L^2(U_i) \cong M_{n+1}(K)$. Thus there is an isometric * homomorphism

$$\psi\circ\tau\colon\ C^*(X_F,F) \to A_n.$$

To show that $\psi\circ\tau$ is onto, one need only apply a C*-algebra version of the Stone-Weierstraus Theorem. Since A_n is type I, it is enough to show that $\mathrm{In}(\psi\circ\tau)$ is a rich subalgebra of A_n. (Cf 11.1.1 and 11.1.4., [D]). The spectrum of $\mathrm{Im}(\psi\circ\tau)$ is given by the family of mutually inequivalent irreducible representations of $C^*(X_F,\ ):\{\pi_{(0,t)}\}_{t>0}$ and $\pi_{(x_i,0)}\}_{i=0,1\ldots,n}$, where $(x_i,0)$ is the middle point of U_i. Here we use the standard notation: for $P \in X_F$, π_p means the representation of $C^*(X_F,F)$ at $L^2(\mathbb{G}_p)$.

If we consider $\mathbb{G}(X_F,F)$ given by (4.7), then the one-densities are just continuous functions multiplying the Lebesgue measure. Thus for $f \in C_c(\mathbb{G}(X_F,F))$ and $\xi \in L^2(\mathbb{R})$, $p \in X_F$, in terms of Euclidean coordinates,

$$\pi_p f\xi(\gamma) = \int_{\gamma\in\mathbb{G}} f(\gamma_1)\xi(\gamma_2)$$

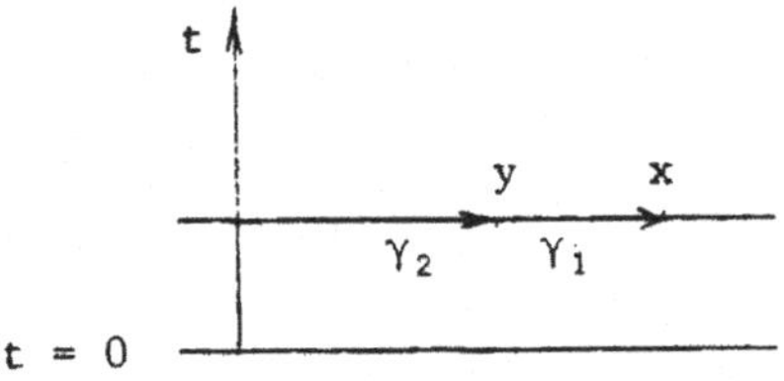

becomes

$$(4.11)\qquad \pi_{(0,t)} f\xi(x) = \int_{\mathbb{R}} f(t,x,y)\xi(y)dy$$

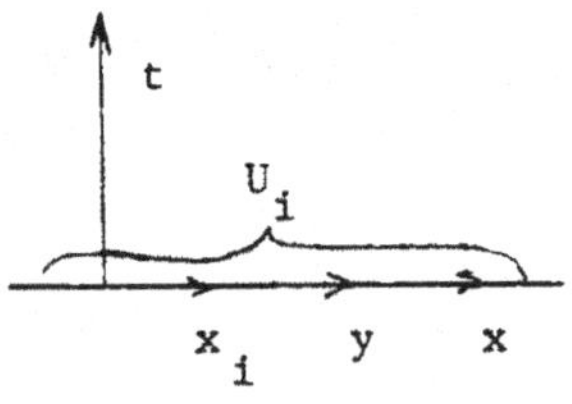

for $t \in 0$, and

$$(4.12)\qquad \pi_{(x_i,0)} f\xi(x) = \int_{U_i} f(0,x,y)\xi(y)\ dy$$

for $i = 0, 1, \ldots, n$. Note that if x_i is chosen differently in U_i, the representation $\pi_{(x_i,0)}$ can be given by the same equation (4.12). Compare (4.11) with (4.6).

Clearly $\{\psi^{-1}\circ\tau^{-1}\circ\pi_{(t,0)}\}_{t>0}$ and $\{\psi^{-1}\circ\tau^{-1}\circ\pi_{(0,x_i)}\}_{i=0,\ldots,n}$ are mutually inequivalent representations of A_n, and exactly representatives of $\hat{A}_n$. Q.E.D.

4.1.3. Note that there is an obvious way to write A_n as an extension of $C_0((0,1))\otimes K$ by K^n. The quotient homomorphism is evaluation of elements of A_n at $t = 0$.

If we drop the assumption in Thm 4.1.2. that the set F is of measure zero, the conclusion is still valid except that we no longer are able to exhibit explicitly a *-isomorphism form $C^*(X_F,\mathcal{F})$ to A_n. However the isomorphism is still canonical with respect to the extension mentioned above.

Theorem 4.1.4. *Under the same assumption as in Thm. 4.1.2., except that we drop the requirements that the set* F *is of measure zero, there is again an exact sequence*

$$0 \to C_0(0,\infty)\otimes K \to C^*(X_F,\mathcal{F}) \to K^{n+1} \to 0. \tag{4.13}$$

As an extension of $C_0(0,\infty)\otimes K$ *by* K^{n+1}, *the* C**-algebra* $C^*(X_F,\mathcal{F})$ *is again isomorphic to* A_n *given by* (4.1) ((4.2)). *Here* $n = 1, 2, \ldots, \infty$.

Proof. We shall use the transversal method and apply the Hilsum-Skandah's stability theorem [H-S].

Again we have $\mathbb{R}\setminus F = \cup_{k=0}^{n} U_k$, where $\{U_k\}$ is a collection of disjoint open intervals. Let $P_k = (x_k,0)$ be the middle point of U_k. Let $T_k = \{(x_k,t)\}_{t\geq 0}$ and $T = \cup_{k=0}^{n} T_k$. Then T is a faithful transversal of (X_F,F).

Consider the holonomy maps between the transversals T_i and T_j in the foliation (X_F,F). We see that an element γ of the discrete groupoid $\mathbf{G}_T^T$ is determined by $(S(\gamma),\ t(\gamma))$, which is of the forms either $((x_i,t),\ (x_j,t))$, for some $t > 0$; or $((x_i,0),\ (x_i,0))$, $i, j = 0, 1, \ldots, n$. We let $(t,i,j) = ((x_i,t),\ (x_j,t)$. Then

$$\mathbf{G}_T^T = \{(t,i,j) \in [0,\infty) \times \mathbb{Z}_{n+1}^2 \mid i = j, \text{ if } t = 0\}.$$

For any $f \in C_c(G_T^T)$, let $f_{i,j}(t) = f(t,i,j)$, $i, j = 0, 1, \ldots, n$. Then $f_{i,i} \in C_c[0,\infty)$ and $f_{i,j} \in C_c(0,\infty)$ if $i \neq j$. Conversely, any such family $(f_{i,j})_{i,j=0,\ldots,m}$, determines an element as long as, for the case $m = \infty$, all but a finite number of its elements are 0 of $C_c(G_T^T)$. Let ϕ be the map sending f to the matrix $(f_{i,j})$. It is clear that ϕ is a bijective linear map from $C_c(G_T^T)$ onto the algebra

$$D = \{(f_{ij})_{(n+1)\times(n+1)} \mid f_{i,j} = 0 \text{ except for a finite number of } i,j\text{'s},\ f_{i,j} \in C\,(0,\infty) \text{ if } i \neq j,\ f_{i,i} \in C_c[0,\infty)\}.$$

Clearly ϕ is a $*$-isomorphism. For let $g \in C_c(G_T^T)$, $t \geq 0$. Then

$$\begin{aligned}(f*g)_{i,j}(t) &= (f*g)((x_i,t),\ (x_j,t))\\ &= \textstyle\sum_{k=0}^{n} f((x_i,t),x_k,t))g((x_k,t),\ (x_j,t))\\ &= \textstyle\sum_{k=0}^{n} f_{i,k}(t)\ g_{k,j}(t).\end{aligned}$$

Thus $\phi(f*G) = \phi f \cdot \phi g$. Also $\phi(f^*) = (\phi f)^*$ can be easily verified.

As subalgebras of $C^*(G_T^T)$ and $C([0,\infty)) \otimes M_n$ resp., both algebras $C_c(G_T^T)$ and D are equipped with the sup norm topology and the isomorphism is obviously an isometry. Therefore ϕ can be extended to an isometry from $C^*(G_T^T)$ onto the completion of D, i.e.,

$$(4.14) \qquad \overline{D} = \{f \in C_0[0,\infty) \otimes M_n(\mathbb{C}) \mid f(0) = \begin{pmatrix} a_0 & & 0 \\ & \ddots & \\ 0 & & a_n \end{pmatrix}, a_i \in \mathbb{C}\}.$$

Now we apply the Hilsum-Skandlis stability theorem and conclude that $C^*(X_F, F)$ and $A_n = \overline{D} \otimes K$ are isomorphic.

Let $\mathring{G}_T^T$ be the interior of G_T^T and ∂G_T^T be the boundary i.e., the $(n+1)$ points $\{(x_i, 0)\}_{i=0,\ldots,n}$. There are obvious commuting exact sequences

$$(4.15) \qquad \begin{array}{ccccccccc} 0 & \to & C^*(\mathring{G}_T^T) & \longrightarrow & C^*(G_T^T) & \longrightarrow & C^*(\partial G_T^T) & \to & 0 \\ & & \downarrow \cong & & \downarrow \cong & & \downarrow \cong & & \\ 0 & \to & C_0(0,\infty) \otimes M_{n+1} & \longrightarrow & \overline{D} & \longrightarrow & \mathbb{C}^{n+1} & \longrightarrow & 0 \end{array}$$

We also have the following commuting exact sequences

$$\begin{array}{ccccccccc} 0 & \longrightarrow & C^*(\mathring{X}_F, F) & \longrightarrow & C^*(X_F, F) & \longrightarrow & C^*(\mathbb{R} \setminus F, F) & \longrightarrow & 0 \\ & & \wr\downarrow & & \wr\downarrow & & \wr\downarrow & & \\ 0 & \longrightarrow & C_0(0,\infty) \otimes M_{n+1}(C) & \longrightarrow & A_n & \longrightarrow & K^{n+1} & \longrightarrow & 0 \end{array}$$

Q.E.D.

Remark 4.1.5. Let F be any closed subset of $\mathbb{R}$. Let V be the maximal open subset of F. We may write $V = \bigcup_{j=1}^{m} U_j$ as a disjoint union of open intervals, $0 \leq m \leq \infty$. From the proof of Thm 4.1.2, we check easily that $(X_F, \mathcal{F})$ is isomorphic to

$$\{f \in C_0([0,1), M_{n+1}(K)) \mid f(0) = \begin{pmatrix} a_0 & a_1 & & \\ & & \ddots & \\ & & & \ddots \end{pmatrix}, a_i \in K, a_{n_j} \equiv 0, j = 1,\dots,m\}$$

where $0 \leq n \leq \infty$ and $M_{n+1}(K)$ stands for $K \otimes K$, if $n = \infty$. Also the n_j-th position on the diagonal corresponds to the interval U_j, for $j = 1, \dots, m$. Applying Thm. 4.1.4., we get another expression for $C^*(X_F, \mathcal{F})$. For instance if $m = 1$ we have shown in particular that the C^*-algebra

$$\{f \in C_0([0,1), M_2(K)) \mid f(0) = \begin{pmatrix} a & \\ & 0 \end{pmatrix}, a \in K\}$$

is isomorphic to $C\,[0,1) \otimes K$. Even in this simple situation, how to construct an explicit isomorphism is not obvious (§5, Part I.).

4.2 C*-algebras of distinguished trees

We shall introduce the concept of "a tree of C*-algebras", which will be useful in describing the structures of noncommutative CW-complexes. To motivate our definition, we recall the concept of "a graph of groups" in combinatorial group theory.

Definition 4.2.1. (p. 37 [Se]). *A graph of groups* (G,Γ) consist of a graph Γ, a group G_x for each x in $\mathrm{vert}(\Gamma)$ and a group G_y for each

y in edge Γ, together with a *monomorphism* $G_y \to T_{t(y)}$ such that $G_y = G_{\bar{y}}$. When Γ is a tree T, then (G/T) is a *tree of groups*.

4.2.2 Let $(G_i)_{i\in I}$ be a family of groups and $F_{i,j}$ be a set of homomorphisms of G_i into G_j for each pair (i,j). Then there exist a group $G = \varinjlim G_i$ and a family of homomorphisms $f_i : G_i \in G$ such that $f_j \circ f = f_i$ for all f in F_{ij} with the following universal property: If H is a group and $h_i : G_i \to H$ is a family of homomorphisms such that $h_j \circ f = h_i$ for all f in F_{ij}, then there is exactly one homomorphism $h : G \to H$ such that $h_i = h \circ f_i$. The group G is the *direct limit* of (G_i, F_{ij}). Clearly G is unique up to isomorphism.

Note that F_{ij} may be empty and that neither f_i nor f_{ij} are assumed to be one to one.

Let (G,T) be a tree of groups. The direct limit $G_T = \varinjlim (G,T)$ is the "amalgam of the G_x along the G_y'. For instance, if T is $\overset{x_1}{\circ}\underset{y}{\rule{3em}{0.4pt}}\overset{x_2}{\circ}$, $G_{x_1} = \mathbb{Z}_6$, $G_x = \mathbb{Z}_4$ and $G_y = \mathbb{Z}_2$ with the inclusion maps, then $G_T = \mathbb{Z}_4 \underset{\mathbb{Z}_2}{*} \mathbb{Z}_6 \cong SL(2,\mathbb{Z})$.

4.2.3 Consider a construction which is in a certain sense dual to the one above for familys of C*-algebras. Let $(A_i)_{i\in I}$ be a (*not* necessarily countable) family of C*-algebras, and for each pair (i,j), let F_{ij} be a set of *-homomorphisms from A_i into A_j (F_{ij} can be empty). If I is *finite*, then there is a C*-algebra A and a family of homomorphisms $\phi_i : A \to A_i$ such that $\phi \circ \phi_i = \phi_j$ for all $\phi \in F_{ij}$ with the following universal property: If B is a C*-algebra and $\psi_i : B \to A_i$ is a family of homomorphisms such that $\phi \circ \psi_i = \psi_j$ for all ϕ in F_{ij}, then there is exactly one homomorphism $\psi : B \to A$ such that $\phi_i \circ \psi = \psi_i$.

To see the existence of such (A, ϕ_i), we simply let A be the subalgebra of $\prod_{i \in I} A_i$ consisting of elements $(a_i)_{i \in I}$, satisfying the relations $\phi(a_i) = a_j$ for all $\phi \in F_{ij}$, and all pair (i,j). Obviously such (A, ϕ_i) is unique up to isomorphism. We always define (A, ϕ_i) in this way, even when I is infinite, and call A the fibred product, or the wedged product, of (A_i, F_{ij}).

__Example 4.2.4.__ Let $p_i : A_i \to A_0$, $i = 1, \ldots, n$, be subjective homomorphisms of C*-algebras, $n < \infty$. Then the corresponding fibred product $A = \bigvee_{A_0} A_i = \{(a_1, \ldots, a_n) \in \bigoplus_{i=1}^n A_i \mid p_i(a_i) = p_j(a_j)\}$, with $\phi_i : A \to A_i$ being the projections. The C*-algebra A can be viewed as the noncommutative locally compact space obtained by glueing the A_i along A .

4.2.5 When the index set I is countably infinite, the fibred product A of (A_i, F_{ij}) is no longer a C*-algebra, but a σ-C*-algebra as defined by Arveson (p. 255 [A]).

By definition a *σ-C*-algebra* is a complex *-algebra A, together with a sequence $\|\cdot\|_1, \|\cdot\|_2$, of C*-seminouns on A satisfying the conditions

(i) $x_n = 0$ for all $n \geq 1 \Rightarrow x = 0$

(ii) A is complete relative to the metric

$$d(x,y) = \sum_{n=1}^{\infty} 2^{-n} \frac{\|x,y\|_n}{1 + \|x-y\|_n}$$

A *projective system* of C*-algebras is a sequence $A_1, A_2, \ldots,$ of C*-algebras with a sequence of surjective *-homomorphisms $\pi_n : A_{i+1} \to A_i$. The fibred product of (A_i, π_i) constructed in 4.2.3. is just the usual inverse limit $A = \varprojlim A_i$ which is a σ-C*-algebra. Conversely, any σ-C*-algebra is the inverse limit of such a projective system (pp. 256-258, [A]). However the following proposition tells us that σ-C*-algebras naturally arise in a much more general context.

Proposition 4.2.6. *Assume the notation in* 4.2.3. *and allow the index set* I *to be countable infinite. Then with the natural* C*-*seminorms* $\|\cdot\|_i$ *inherited from* A_i *the fibred product* A *is a* σ-C*-*algebra. Moreover* (A, ϕ_i) *has the following universal property: If* B *is a* σ-C*-*algebra and* $\psi_i : B \to A_i$ *is a family of continuous* *-*homomorphisms such that* $\phi \circ \psi_i = \psi_j$ *for all* ϕ *in* F_{ij}, *then there is exactly one continuous homomorphism* $\psi : B \to A$ *such that* $\phi_i \circ \psi = \psi_i$.

Proof. Axiom (i) is obvious.

Let $\{x_n\}$, $x_n = (a_i^n)$, be a Cauchy sequence in A. Then $\{A_i^n\}$ is also a Cauchy sequence in A_i for each fixed i. So $a_i^n \to a_i$. Since $\phi(a_i^n) = a_j^n$ for all $\phi \in F_{ij}$, we have $\phi(a_i) = a_j$. Thus $x = (a_i)$ is in A.

The rest of the discussion is also routine, and so is omitted. Q.E.D.

Definition 4.2.7. *A regular* $\mathbb{R}$-*tree of* C*-*algebras*, or just a *tree of* C*-*algebras*, (T,A,ϕ) consists of a regular $\mathbb{R}$-tree (T,d,V,E), a C*-algebra A_x for each x in V. A C*-algebra, A_y for each $y \in E$, and a C*-

algebra A_z for each $z \in V'$ such that

i) If $\gamma(y) = (x_1, x_2) \in V \times \overline{V}$ for some $y \in E_{x_1}$, then there are surjective homomorphisms $\phi_{x_i y}^{\circ\circ} A_{x_i} \to A_y$, $i = 1,2$. Moreover if $x_2 \in V$ then $A_y = A_{\overline{y}}$, if $x_2 = z \in V'$ then $A_{x_2} = A_z = A_z$ and $\phi_{x_2,y} = \text{id}$.

ii) If $x_n \to z$, $x_n \in V$, $z \in V'$, then A_{x_n} is eventually a continuous field of C*-algebras with fibers isomorphic to A_z, and for any sequence $\{y_n\}$, $y_n \in E_{x_n}$, A_{y_n} is eventually isomorphic to A_z.

Here by "eventually" for the elements in a sequence, we mean "for all but finitely many".

4.2.8. In the next section, 4.2.9., we shall define a tree of C*-algebras for a distinguished tree. In order to motivate our definition, we now assume that the distinguished tree T is associated to a foliation $(\Omega,\mathcal{F})$ (Thm. 3.5.11.) and describe explicitly the C*-algebras of various types of saturated subregion of Ω and carefully investigate how they are related in terms of the structure of the tree $T(\mathcal{F})$. The discussion in this section will actually be more detailed than just motivation, in fact it will serve as part of the proof of Thm. 4.3.1.

Let $T = T(\mathcal{F}) = (T,d,V,E)$ be a distinugished tree as usual. Recall from Def. 3.2.9. that a vertex $x \in V$ is a center of $\Delta(0)$, for some canonical region 0 (Fig. 3.2.). We associate to x a closed saturated subset Ω_x of Ω according to the following two possibilities.

(a) $\Delta(0)$ is of type (iv) (Cor. 1.5.8.). Thus x is the center of one of the two polygonoids contained in $\Delta(0)$, say, P_x. Let L be the leaf in 0 dividing the area of 0 in half. Then Ω_x is the closed region in Ω bounded by L and the $\Delta^{-1}(S_i)$, where $S_1, S_2, \ldots,$ are the closed sides of P_x (see (iv), Fig. 3.2). Clearly Ω_x is conjugate to the foliated region shown in (b), Fig. 3.6., where L is a nonseparatrix.

(b) $\Delta(0)$ is not of type (iv). Thus x is the center of $\Delta(0)$. Let Ω_x be the closure of the canonical region 0 in Ω, i.e., the union of 0 and all the separatrices boundary it. We easily recognize that if $\deg x = 1$, then Ω_x is conjugate to a foliated region of the type shown in (a), Fig. 3.6., while if $\deg x = 0$,, then Ω_x is conjugate to that as shown in (b), Fig. 3.6, with $L = L_y$ being a separatrix.

In either case each $y_i \in E_x$ with $\deg y_i = 0$ corresponds to a closed side of the region $\Delta(\Omega_x)$. By Thm 4.1.2., there is a isomorphism ϕ_x from the C*-algebra $C^*(F_x, F)$ of the foliation to the C*-algebra A_x, where A_x is defined as follows depending on deg x:

1) If $\deg x = 0$ and $\mathrm{Inc}(x) = n + 1$, then $A_x = \overline{A}_n$ where

$$\overline{A}_n = \{a \in C([0,1], M_n(K)) \,|\, a(0) = \mathrm{diag}(a_1, \ldots, a_n),\ a_i \in K\}.$$

This is a "1-cell, one end of which has n-branches" (and n can be ∞).

2) If $\deg x = 1$ and $\mathrm{Inc}(x) = n$, then $A_x = A_n$ where

$$A_n = \{a \in C([0,1), M_n(K)) \,|\, a(0) = \mathrm{diag}(a_1, \ldots, a_n),\ a_i \in K\}.$$

This is a "1-cell with one end open and with n-branches at the other end".

The inclusion of each separatrix L_y into Ω_x induces a surjective homomorphism $\gamma_{x,y}$ from $C^*(\Omega_x, F)$ onto $C^*(L_y, F)$, for each $y \in E_x$, see Prop. 1.5., Part I. Here the quotient map $\gamma_{x,y}$ is just the completion of the restriction map from $C_c(G(\Omega_x,F)$ onto $C_c(L_y,F))$. Recalling our convention (Rem. 3.3.2.), for any $x \in V_0$ we always enumerate $E_x = \{y_0, y_1, y_2, \dots\}$, where $\deg y_0 = 1$ if $x \in V_0 \cap B_T$. For $x \in V_1$, we enumerate $E_x = \{y_0, y_2 \dots\}$. In either case, only y_0 and y_1 may lie on some Γ_z. Depending on the $\mathbb{Z}_2$-grading of an edge $y \in E_x$ (regardless deg x), we have the following two cases.

i) $\deg y = 0$. Say $y = y_i$, $i = 1,2, \dots$. Then we have a commuting diagram as shown in Diag. 4.1. There the isomorphism ϕ_x from $C^*(\Omega_x,F)$ onto A_x is constructed as in Thm. 4.1.2.,

$$\begin{array}{ccc} C^*(\Omega_x,F) & \xrightarrow{\gamma_{x,y_i}} & C^*(L_{y_i},F) \\ \phi_x \downarrow \simeq & & \phi_{y_i} \downarrow \simeq \\ A_x & \xrightarrow{f_{x,y_i}} & A_{y_i} = k \end{array}$$

Diagram 4.1.

$$\begin{array}{cccc} C^*(\Omega_x,F) & & \xrightarrow{\gamma_{x,y}} & C^*(X_y,F) \\ \simeq \downarrow \phi_x & & & \phi_y \downarrow \simeq \\ A_x & \xrightarrow{f_{x,y}} & M_n(K) \xrightarrow[\simeq]{\sigma_{n,1}(x)} & K \end{array}$$

Diagram 4.2.

the evaluation f_{x,y_i} sends a to a_i, for $a \in A_x$ given as in 1) or 2) above, and the isomorphism ϕ_{y_i} is just the quotient of the isomorphism ϕ_x.

ii) $\deg y = 1$. Thus $y = y_0$ and $x \in B_T \cap V_0$. Recall that y corresponds to the open side of P_x. Corresponding to y there is a leaf L_y which may be (when $\gamma(y) = (x,x')$, with $x' \in V$ and $\deg \bar{y} = 0$, or with $x' = z \in V'$) or may not be (when $\deg \bar{y} = 1$) a separatrix. In this case we have a commuting diagram as shown in Diag. 4.2. There the homomorphisms ϕ_x, $r_{x,y}$, and ϕ_y are defined as above, while $f_{x,y}$ is the evaluation map taking a to $a(1)$, for $a \in A_x$ defined as in 2). The isomorphism $\sigma_{n,1}(x)$ is induced by the isomorphism of Hilbert spaces $\bigoplus_{i=1}^{n} L^2(U_i) \simeq L^2(\mathbb{R})$, where the decomposition of $\mathbb{R}$ into disjoint union of intervals U_i was made in the proof of Thm. 4.1.2.

For each $x \in V_0$, with $\mathrm{Inc}(x) = n+1$, we associate to x an isomorphism $\sigma(x) = \sigma_{n,1}(x)$ from $M_n(K)$ onto K, such that

(α) if $n = 1$, then $\sigma(x) = \mathrm{Id}$.

(β) if $\gamma(y) = (x,x')$, such that $x' \in V$ and $\deg y = \deg \bar{y} = 1$, then $\sigma(x')^{-1}\sigma(x)$ is an isomorphism from $M_n(K)$ onto $M_{n'}(K)$, where $n' = \mathrm{Inc}(x') - 1$, with the property that the restriction of $\sigma(x')^{-1}\sigma(\bar{x})$ to $e_{1,1} \otimes K$ is the identity. Here $e_{1,1} \in M_n(x)$.

(γ) if $\{x_m\}$ is a sequence in V_0 converging to some point in V', then $\|\sigma(x_m)(e_{1,1} \otimes T) - T\| \to 0$ for any $T \in K$. Here $e_{1,1} \in M_{n_m}(\mathbb{C})$, and $n_m = \mathrm{Inc}(x_m)-1$.

Some comments about the conditions (β) and (γ) are in order. Note first that any isomorphism from $M_n(K)$ to $M_{n'}(K)$ must be spatial, i.e., it arises from an isomorphism of corresponding Hilbert spaces. In order for the condition (β) to be true, we choose the decomposition of Hilbert spaces $L^2(\mathbb{R}) \simeq \bigoplus_{i=1}^{n} L^2(U_i)$ and $L^2(\mathbb{R}) \simeq \bigoplus_{i=1}^{n} L(U_i')$ such that $U_1 = U_1'$. For the geometric meaning at the level of foliations, see Fig. 4.3.

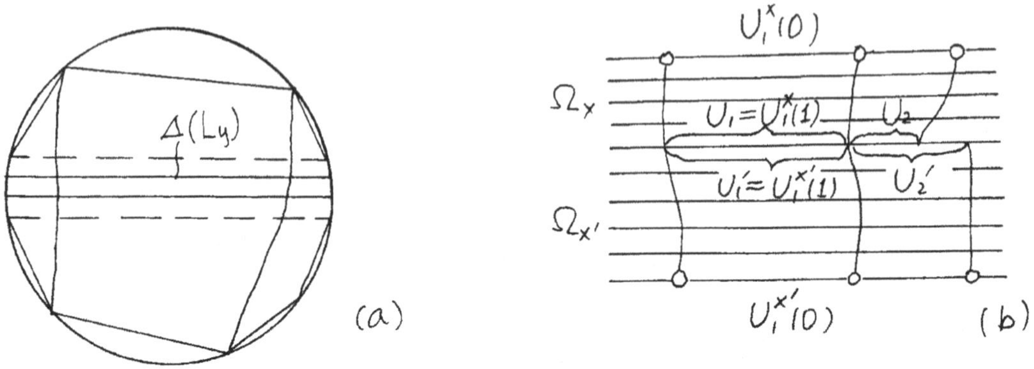

Figure 4.3.

In order to see how Condition (γ) can be satisfied in the string of isomorphisms: $M_{n_m}(K) \simeq K(\bigoplus_{i=1}^{n_m} L^2(U_i^m)) \simeq K(L^2(\mathbb{R})) \simeq K$, we choose the sequences of intervals $U_1^m \nearrow \mathbb{R}$ as $m \to \infty$. This is indeed the motivation while we consider the properties of C*-algebras of foliations, when limit separatrices do occur. See Fig. 4.3.c)

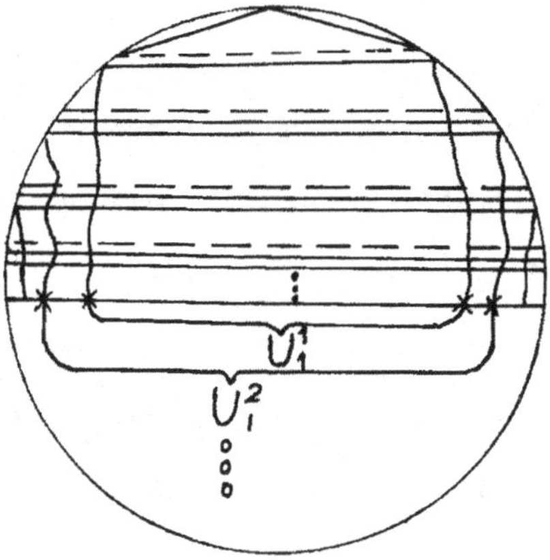

Figure 4.3.c)

Definition 4.2.9. For a distinguished tree $T = (T,d,V,E)$, a regular $\mathbb{R}$-tree of C*-algebras (T,A,ϕ) is defined as follows: for each $y \in E$ and each $z \in V'$ we let $A_y = K$ and $A_z = K$; for each $x \in V$, the C*-algebra A_x is defined in 1) or 2), 4.2.8., depending on deg(x).

For each $y \in E_x$, for all $x \in V$, the surjective homomorphism $\phi_{x,y}: A_x \to A_y$ is defined as follows. For $a \in A_x$, with $a(0) = \mathrm{diag}(a_1, \ldots, a_n)$ (recall i) and ii), 4.2.8.).

i) if $y = y_i$, $i = 1, 2, \ldots,$ let $\phi_{x,y}a = a_i$;

ii) if $y = y_0$, let $\phi_{x,y}a = \sigma_{n,1}(x)a(1)$

where the set $\sigma = \{\sigma_{n,1}(x)\}_{x \in V}$ was defined in 4.2.8.

4.2.10. Note that the set $\phi = \{\phi_{x,y}\}$ is determined by $\sigma = \{\sigma(x)\}$. Thus we may also write (T,A,σ) for (T,A,ϕ). Throughout the remainder of this paper, we shall be exclusively concerned with this special class of distinguished trees of C*-algebras, defined in Def. 4.2.9., and we shall simply call them as *trees of C*-algebras*.

Definition 4.2.11. Let (T,A,σ) be a tree of C*-algebras defined as in Def. 4.2.9. The *global* fibered product of (T,A,σ) written as $C^*(T,A,\sigma)$, consists of all $(a_x) \in \prod_{x \in \overline{V}} A_x$ satisfying the following conditions.

(1) (local glueing conditions) i) let $x, x' \in V$ be adjacent, $\gamma(y) = (x,x')$, $\gamma(\overline{y}) = (x',x)$ for a $y \in E$; then $\phi_{x,y}a_x = \phi_{x',y'}a_{x'}$.
ii) if for some $y \in E$, $\gamma(y) = (x,z)$, $x \in V$, $z \in V'$, then $\phi_{x,y}a_x = a_z$.

(2) (global glueing conditions). Let $\{x_m\}$ be a sequence in V converging to $z \in V'$. Then

$$\sup_{t\in[0,1]} \|a_{x_m}(t) - e_{1,1} \otimes a_z\| \to 0 \quad \text{as} \quad m \to \infty$$

(3) (vanishing at ∞) if a sequence $\{x_m\}$ in $\overline{V}$ converges to ∞, in the sense that any compact subset of T contains only finitely many x_m, then $\|a_{x_m}\| \to 0$ as $m \to \infty$.

4.2.12. Note that if $z_m \to z_0$, $z_m \in V'$, $i = 0, 1, \ldots,$ then the global glueing condition in 4.2.11 forces $\|a_{z_m} - a_{z_0}\| \to 0$. It is easy to check that $C^*(T,A,\sigma)$ is a separable C^*-algebra.

The three conditions in Def. 4.2.11 need some explanation. The condition (3) of vanishing at ∞ is quite natural, because $C^*(\Omega,F)$ is just the norm closure of $C_c(G(\Omega,F))$, and for a compact subset E of Ω, there is a compact subset K' of $T(F)$ such that if $L \cap E \neq \phi$, then $\Delta(L)$ is a chord which is either a limit chord or contained in some region in $K(F)$ such that in both cases the associated point (or points, if L is a separatrix, see §3.1) in $T(F)$ is contained in K'. In other words, the "support" in $T(F)$ of an element in $C_c(\Omega,F)$ is again compact.

Note that in Def. 4.2.9, we attached a C^*-algebra A_y to an edge $y \in E$ which somehow disappeared in the explicit expression of the global fibred product in Def. 4.2.11. The reason that they were omitted in the product is that for each element $a \in C^*(T,A,\sigma)$, the value a_y of a at $y \in E_x$ is given by $\phi_{x,y}a_x$, which is $\phi_{x',\overline{y}}a_{x'}$ if $\gamma(y) = (x,x')$, and $x' \in V$ (local glueing condition). For the same reason by the global glueing condition (2), we could have omitted V' from the index set for the global product. We have retained it because of the notational convenience in later discussion.

From the discussion in 4.2.8., the motivation for the local glueing condition (1) is apparent. For any separatrix L_y, with $\gamma(y) = (x,x')$, one requires surely that both the restriction map γ_x from $C^*(\Omega_x,F)$ to $C^*(L_y,F)$ and the restriction map $\gamma_{x'}$ from $C^*(\Omega_{x'},F)$ to $C^*(L_y,F)$ when applied to a_x and $a_x{}'$ respectively, for any $a \in C^*(\Omega,F)$, yield the same element a_y. However, the geometric meaning of the global glueing condition (2) is less apparent and is more technical to derive from the local property of foliations at a neighborhood of a limit separatrix, which is represented by the axiom (5) of Definition 3.3.1., We postpone the discussion of this until the proof of Thm. 4.3.1. At this moment we only point out that there is an equivalent formulation for the global glueing condition (2).

Proposition 4.2.13. *Let* (T,A,σ) *be a tree of* C*-*algebras (Def.* 4.2.9). *Then the condition* (2) *in Def.* 4.2.11 *is equivalent to*

(2') *For any sequence* $\{x_m\}$ *in* V *converging to* $z \in V'$,

$$\sup_{t\in[0,1]} \|\sigma(x_m)(a_{x_m}(t)) - a_z\| \to 0, \quad \text{as} \quad m \to \infty.$$

Proof. The $\sigma(x_m)$ are all isometric as they are isomorphisms of C*-algebras. The condition (2), Def. 4.2.11 is then equivalent to:

(2") $\sup_t \|\sigma(x_m)(a_{x_m}(t)) - \sigma(x_m)(e_{1,1} \otimes a_z)\| \to 0$

as $m \to \infty$. From the axiom (γ) of (3), Def. 4.2.9., we have

$$(\gamma') \qquad \|\sigma(x_m)(e_{1,1} \otimes a_z) - a_z\| \to 0, \; m \to \infty.$$

Combining this with (2"), we get (2'). Conversely (2") can be derived from (2') and (γ'). Q.E.D.

Remark 4.2.14. The condition (a) of Def. 4.2.11 can be phrased still another way. Let $P_1 = \mathrm{diag}(I,0,\ldots,0)$ in $M_n(B(H))$. Then the condition (2) is equivalent to the following two conditions:

$$(2)_1: \qquad \sup_{t\in[0,1]} \|a_{x_m}(t) - p_1 a_{x_m}(t) p_1\| \to 0,$$

and

$$(2)_2: \qquad \sup_{t\in[0,1]} \|p_1 a_{x_m}(t) p_1 - a_z\| \to 0,$$

as $m \to \infty$, when $x_m \to z$. The proof is quite obvious. If we write $p_{1,1}$ for the projection from $M_n(K)$ to K sending a $n \times n$ matrix to the entry in its upper left corner, then $p_1 a_{x_m}(t) p_1 = e_{1,1} \otimes p_{1,1}(a_{x_m}(t))$.

It turns out that the isomorphism class of $C^*(T,A,\sigma)$ is independent of a particular choice of the family $\sigma = \{\sigma_{n_{x,1}}(x)\}_{x\in V}$ of isomorphisms. This will be shown "locally" as well as "globally". Thus the C*-algebra $C^*(T,A,\sigma)$ is inherent to the distinguished tree T. Geometrically, this means that the "noncommutative CW 1-complex" so obtained is independent of the particular way of glueing each pair of "branches"; at each individual joint one may add an arbitrary "twist", as long as, when the vertices converge the "twists" converge too.

The following lemma will furnish the inductive step for the "local" argument.

Lemma 4.2.15 (a) *Let* $S_i = \{A_i, M_n(K), A_x, M_m(K); f^i, f_{x,y}, \sigma_i\}$, $i = 1,2$, *be two families of C*-algebras as discussed in 4.2.3., where we fixed the* *homomoprhisms* $f^i\colon A_i \longrightarrow M_n(K)$, $n \geq 1$, $f_{x,y}\colon A_x \longrightarrow M_m(K)$, *and* $\sigma_i\colon M_n(K) \xrightarrow{\cong} M_m(K)$, $i = 1,2$, $m \geq 2$ *(see Fig. 4.4). Suppose that there is an isomorphism* $\phi_0\colon A_1 \longrightarrow A_2$ *such that* $f_y^2 \circ \phi_0 = f_y^1$, *that* $A_x = \bar{A}_m$, *and* $f_{x,y}$ *is as defined in 4.2.8. Then* ϕ_0 *can be extended to an isomorphism* ϕ *between the global fibred products* $A_{1,x}$ *and* $A_{2,x}$ *such that:*

(1) *for any "boundary evaluation map"* $f_{x,y'}$ *from* A_x *onto* K *(*$\deg y' = 0$*) we have* $f_{x,y'} \circ \phi = f_{x,y'}$*;*

(2) *the restriction of* $\phi|_{A_x}$ *to* $C_0([0,1), e_{1,1} \otimes K)$ *is the identity map.*

(b) *Except for condition (2) the assertion in* (a) *is also true in the case that* $m = 1$ *(*$\deg y = 0$*) so that the homomorphisms are* $f_{x,y}\colon A_x \longrightarrow K$ *and* $\sigma_i\colon M_n(K) \longrightarrow K$ *while now* $\mathrm{rang}(f_{x,y'})$ *may be* $M_\ell(K)$, *where corresponding* A_x *is either* $\bar{A}_\ell$ *or* A_ℓ.

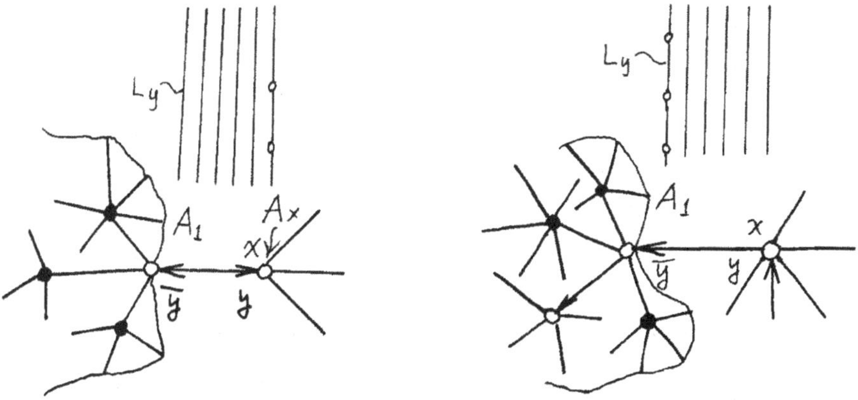

(a) $\deg y = 0$ or 1, but if $y' \in E_x \backslash \{y\}$, then $\deg y' = 0$, $\deg x = 0$.

(b) $\deg y = 0$. There is possibly $y' \in E_x \backslash \{y\}$, with $\deg y' = 1$, $\deg x = 0$ or 1.

In both cases $\deg \bar{y} = 0$ (when $n = 1$), or 1 (when $n > 1$).

Fig. 4.4

Proof. (a) The global fibered product $A_{i,x}$ consists of pairs $(a_i,a_x) \in A_i \oplus A_x$ with $\sigma_i \circ f^i(a_i) = f_{x,y}(a_x)$, $i = 1,2$.

Consider $\sigma_2 \circ \sigma_1^{-1} \in \mathrm{Aut}(M_m(K))$. There is a unitary $U_1 \in M_m(\mathcal{L}(H))$ such that $\mathrm{Ad}_{U_1} = \sigma_2 \circ \sigma_1^{-1}$ and a path U_t, $t \in [0,1]$ of unitaries connecting $U_0 = I$ to U_1. Since $\mathrm{Ad}_{U_1}|e_{1,1}\otimes K = \mathrm{id}$, we may choose $\mathrm{Ad}_{U_t}|e_{1,1}\otimes K = \mathrm{id}$, for all t.

For $(a_1,a_x) \in A_{1,x}$, let $\phi(a_1,a_x) = (\phi_0(a_1),\mathrm{Ad}_{U(\cdot)}f_x)$. Then $\sigma_2 \circ f^2(\phi_0(a_1)) = \sigma_2 \circ f^1(a_1) = (\sigma_2 \circ \sigma_1^{-1})(\sigma_1 f^1(a_1)) = \mathrm{Ad}_{U_1}(f_{x,y}(a_x)) =$
$= f_{x,y}(\mathrm{Ad}_{U(\cdot)}(a_x))$, hence $(\phi_0(a_1),\mathrm{Ad}_{U(\cdot)}(a_x)) \in A_{2,x}$. One verifies easily that ϕ is an isomorphism from $A_{1,x}$ onto $A_{2,x}$. Assume $y' \in E_x$, with $\deg y' = 0$; hence y is one of the y_i (see 4.2.8(i)). Since $\mathrm{Ad}_{U_0} = I$, it follows that $f_{x,y'}(\mathrm{Ad}_{U_0}f_x) = f_{x,y'}(f_x)$ and $f_{x,y'} \circ \phi = f_{x,y'}$ for any $y' \in E_x \setminus \{y\}$.

Conclusion (2) follows from our choice that $\mathrm{Ad}_{U_t}|e_{1,1}\otimes K = \mathrm{id}$, for all t.

(b) Let $\mathrm{Ad}_{U_1} = \sigma_2 \circ \sigma_1^{-1} \in \mathrm{Aut}(K)$, where $u_1 \in \mathcal{U}(H)$. Then there is a path U_t of unitaries in $M_\ell(\mathcal{L}(H))$ such that $U_0 = \mathrm{diag}(1,\ldots,1,U_1,1,\ldots,1)$, where the i-th position of U_1 corresponding to the $y = y_i$ and $U_1 = I$. Again we let $\phi(a_1,a_x) = (\phi_0(a_1),\mathrm{Ad}_{U_t}a_x)$. The rest of the proof is completely analogous to that of (a). Q.E.D.

Theorem 4.2.16. *Let* $T = (T,d,V,E)$ *be a distinguished tree such that the set* V *of vertices is closed. Then up to isomorphism the global fibred product* $C^*(T,A,\sigma)$ *is independent of a particular choice of the set* $\sigma = \{\sigma(x)\}$ *in 4.2.8.*

Proof. Since V is a closed subset of T, we have $V' = \emptyset$ and $K \cap V$ is finite for every compact subset K of T. There exists a sequence $\{(T_k, d, V_k)\}$ of compact regular $\mathbb{R}$-subtrees of (T,d,V) such that $V_k = T_k \cap V$, $T_k \subset T_{k+1}$, and $\bigcup_{k=1}^{\infty} T_k = T$.

Suppose $\sigma_i = \{\sigma^i_{n_x,1}(x)\}$, $i = 1,2$, are two sets of isomorphisms as in Definition 4.2.9. Let $C^*(T,A,\sigma^i)$, $i = 1,2$, be the two corresponding global fibred products. For each $k = 1,2,\ldots$, let I^i_k consist of the elements $(a_x)_{x\in V_k}$ such that $(a_x)_{x\in V} \in C^*(T,A,\sigma^i)$ and $f_{x,y}(a_x) = 0$ if $\gamma(y) = (x,x')$ for $x' \notin V_k$, $i = 1,2$. Then one verifies that I^i_k is an ideal of $C^*(T,A,\sigma^i)$ if we identify $(a_x)_{x\in V_k}$ with $(\tilde{a}_x)_{x\in V}$ where $\tilde{a}_x = a_x$ if $x \in V_k$, $\tilde{a}_x = 0$ if $x \notin V_k$.

By Lemma 4.2.15, there is an isometric isomorphism ϕ_k from I^1_k onto I^2_k such that ϕ_{k+1} is an extension of ϕ_k, $k = 1,2,\ldots$. Now by axiom (3) of Definition 4.2.11, $I^i = \bigcup_{k=1}^{\infty} I^i_k$ is dense in $C^*(T,A,\sigma^i)$, $i = 1,2$. We can extend the isometry $\cup\phi_k : I^1 \to I^2$ to an isomorphism ϕ from $C^*(T,A,\sigma^1)$ onto $C^*(T,A,\sigma^2)$. Q.E.D.

Lemma 4.2.17. *Let* $u_m \in U(H)$ *be a sequence of unitary operators such that for any* $T \in \mathcal{L}(H)$ *one has* $\|\mathrm{Ad}\, U_m(T) - T\| \to 0$ *as* $m \to \infty$. *Suppose* H_m *are Hermitian operators determined by* $U_m = e^{2\pi_i H_m}$. *Denote* $U^t_m = e^{2\pi_i tH_m}$ *for* $0 \le t \le 1$. *Then*

$$\sup_{t\in[0,1]} \|\mathrm{Ad}\, U^t_m(T) - T\| \to 0 \quad as \quad m \to \infty . \tag{4.16}$$

Proof. Fix $T \in \mathcal{L}(H)$. By assumption

$$\|TU_m - U_m T\| \to 0 . \tag{4.17}$$

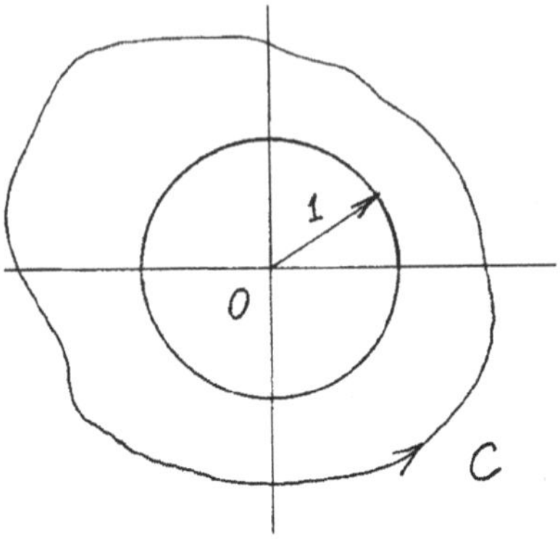

Let C be a simple closed curve encompassing the unit circle in the complex plane which contains, of course, $\mathrm{spec}(U_m)$ for all m. Let $\ell_0 = \inf_{z \in C} |z|$ and $\ell_1 = \sup_{z\in C} |z|$. Then $1 < \ell_0 \leq \ell_1 < \infty$. Given any $\varepsilon > 0$, by (4.17) there is $M > 0$ such that $\|TU_m - U_m T\| < \varepsilon$ when $m > M$, then $\|T(z-U_m) - (z-U_m)T\| < \varepsilon$ for $z \in C$. Thus

$$(4.18) \qquad \|(z-U_m)^{-1}T - T(z-U_m)^{-1}\| < \varepsilon\|(z-U_m)\|^{-2} \leq \varepsilon(\ell_0-1)^{-2} .$$

Applying continuous functional calculus

$$TU_m^t - U_m^t T = \frac{1}{2\pi i}\oint_C \{T(z-U_m)^{-1} - (z-U_m)^{-1}T\} z^t dz ,$$

for $t \in [0,1]$. Thus

$$\begin{aligned} \|\mathrm{Ad}\, U_m^t(T) - T\| &= \|TU_m^{-1} - U_m^t T\| \\ &\leq \frac{1}{2\pi}\oint_C \|T(z-U_m)^{-1} - (z-U_m)^{-1}T\| \, |z|^t ds \\ &\leq \frac{\varepsilon}{2\pi(\ell_0-1)} \cdot \ell_1 \cdot |C| \qquad \text{for } 0 \leq t \leq 1 \end{aligned}$$

where $|C|$ is the total length of the curve C. Thus (4.16) holds.

Q.E.D.

Let (T,A,σ^i) be two trees of C^*-algebras where $\sigma^i = \{\sigma^{i(x)}\}$, $i = 1,2$, are two sets of isomorphisms as defined in 4.28. Let T_0

be any sub $\mathbb{R}$-tree of T. For every $a = (a_x)_{x\in V}$ in $C^*(T,A,\sigma^i)$, we define $\psi_{T_0}(a) = (a_x)_{x\in\Gamma\cap\bar{V}}$. Then with the sup norm $\psi_{T_0}(C^*(T,A,\sigma^i))$ is a C*-algebra and ψ_{T_0} is a quotient map. Let $A^i_{T_0} = \psi_{T_0}(C^*(T,A,\sigma^i))$, for $i = 1,2$.

Lemma 4.2.18. *Let* $\Gamma = \Gamma_z$ *be a closed segment as fixed in Remark 3.3.2 and* A^i_Γ, *for* $i = 1,2$, *are two C*-algebras as defined as above. Then there is an isomorphism* $\phi_\Gamma\colon A^1_\Gamma \longrightarrow A^2_\Gamma$ *such that* $f_{x,y} = f_{x,y}\circ\phi_\Gamma$ *for any* $y \in E_x$ *not lying in* Γ, *for all* $x \in V\cap\Gamma$.

Proof. Recall that $\Gamma\cap V_1 = \emptyset$. We consider first the case that $\Gamma\cap V_1 = \emptyset$. Let x be any point in $\mathring{\Gamma}\cap V$. Then $\deg x = 0$ and $A_x = \bar{A}_n$ for some $n = n_x \geq 1$ (4.2.8).

(1) Let $n = 1$. Then $\sigma_{1,1}(x) = I \in \mathrm{Aut}(K)$ by axiom (α), (3) in 4.2.8. We define $\phi_x \in \mathrm{Aut}(A_x)$ by simply letting $\phi_x = I$.

(2) Suppose $n \geq 2$, i.e., $x \in \Gamma\cap B_T\cap V_0$. Recall Remark 3.3.2 and set $E_x = \{y_0,y_1,y_2,\dots,y_n\}$ such that y_0 and y_1 lie on Γ, while $\deg y_0 = 1$. Recall the expression for $A_x = \bar{A}_n$ in (1), 4.2.8. The y_i, for $i \geq 1$ correspond to the end $t = 0$ while y_0 corresponds to the end $t = 1$. We now define $\phi_x \in \mathrm{Aut}(A_x)$ which "twists" the $t = 1$ end of A_x by $\sigma_2(x)^{-1}\sigma_1(x) = \mathrm{Ad}\, U(x)$, where $U(x) \in \mathcal{U}(\overset{n}{\oplus} H)$. If H is the Hermitian operator given by $U(x) = e^{2\pi i H}$, then we set $U(x)^t = e^{2\pi i tH}$. Thus $U(x)^t$, $0 \leq t \leq 1$, is a norm continuous path in $\mathcal{U}(\overset{n}{\oplus} H)$ connecting $U_x(0) = I$ and $U_x(1)$.

As in Lemma 4.2.15, we define $\phi_x \in \mathrm{Aut}(A_x)$ by $\phi_x(a_x)(t) = \mathrm{Ad}\, U(x)^t a_x(t)$ for $t \in [0,1]$. Then $f_{x,y_0}\circ\phi_x(a_x) = \sigma_2(x)^{-1}\sigma_1(x)a_x(1)$ and $\phi_{x,y_i}\phi_x(a_x) = \phi_{x,y_i}(a_x)$, for $i = 1,2,\dots$, where

$\phi_{x,y_i} = f_{x,y_i}$, $i = 0,1,2,\ldots$ are the boundary evaluation maps as defined in Definition 4.2.9. Now we define $\phi_\Gamma((a_x)_{x\in\Gamma\cap\bar{W}}) = (a'_x)_{x\in\Gamma\cap\bar{W}}$, for any $(a_x)_{x\in\Gamma\cap\bar{W}} \in A^1_\Gamma$, where $a'_x = \phi_x(a_x)$ if $x \in \Gamma\cap V$ and $a'_x = a_x$ if $x \in \Gamma\cap V'$. We claim that $(a'_x)_{x\in\Gamma\cap\bar{W}} \in A^2_\Gamma$. This needs to be shown locally as well as globally, although the vanishing at infinity condition is obviously satisfied for $(a'_x)_{x\in\Gamma\cap\bar{W}}$.

First we check the local condition (1) of Definition 4.2.11. Let $y \in E_x$ and $\gamma(y) = (x,x')$. Depending on $\deg y$ and $\deg \bar{y}$ there are altogether the following three essentially different situations:

Case (a). $\deg y = 1$, $\deg \bar{y} = 0$.

By Axiom (iv) of Definition 3.3.1, we have $\mathrm{Inc}(x) > 2$. So $x' \in V_0\cap\Gamma\cap B_T$. See Fig. 4.5. We find

$$\phi_{x,y}(a'_x) = \sigma_2(x)\left(\sigma_2(x)^{-1}\sigma_1(x)a_x(1)\right) = \sigma_1(x)a_x(1)$$

and

$$\phi_{x',\bar{y}}(a'_{x'}) = f_{x',\bar{y}}(a'_{x'}) = f_{x',\bar{y}}(a_{x'}) = \sigma_1(x)a_x(1) \ .$$

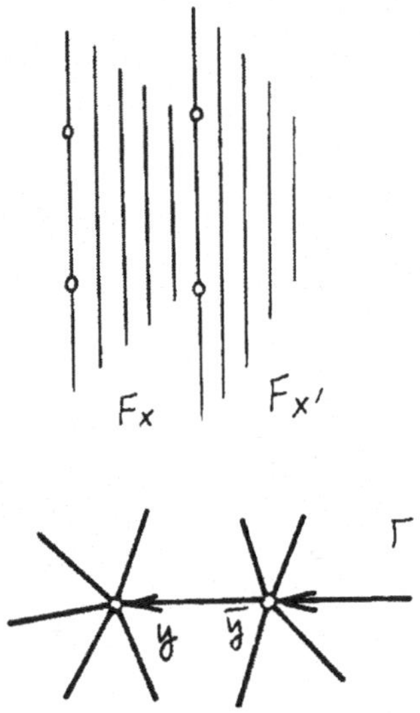

Fig. 4.5

Case (b). deg $y = 1$ and deg $\bar{y} = 1$.

See Fig. 4.6. In this case we have

$$\phi_{x,y}(a'_x) = \sigma_1(x)a_x(1)$$

and

$$\phi_{x',\bar{y}}(a'_{x'}) = \sigma_1(x')a_{x'}(1) .$$

Again by the local glueing condition for $(a_x)_{x\in\bar{V}}$, we have

$$\sigma_1(x)a_x(1) = \sigma_1(x')a_{x'}(1) .$$

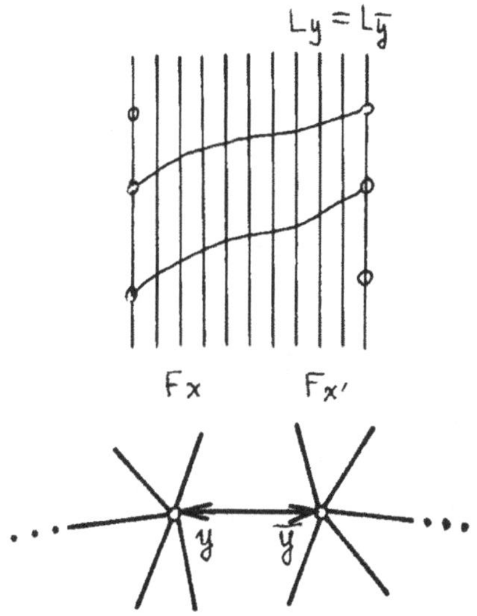

Fig. 4.6

Case (c). deg y = deg $\bar{y} = 0$.

See Fig. 4.7. We have

$$\phi_{x,y}(a'_x) = f_{x,y}(a'_x) = f_{x,y}(a_x) = f_{x',\bar{y}}(a'_{x'}) = \phi_{x',\bar{y}}(a'_{x'}) .$$

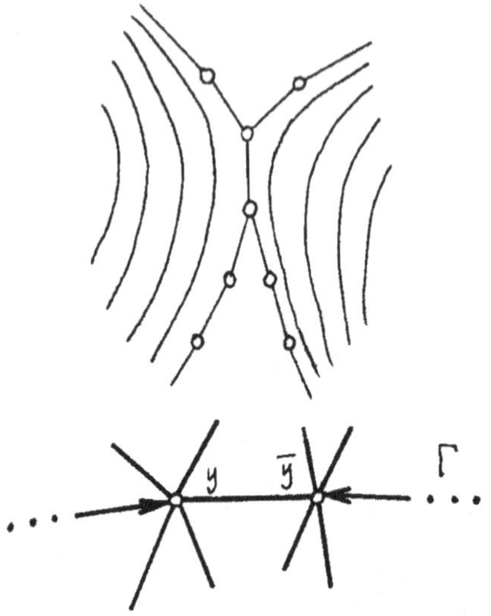

Fig. 4.7

Next we show that the global condition (2) of Definition 4.2.11 holds for $(a'_x)_{x\in\Gamma\cap\bar{V}}$. So let $z \in \Gamma\cap V'$ and $\{x_m\}$ be a sequence in V converging to z. Since $(a_x)_{x\in\bar{V}} \in C^*(T,A,\sigma)$, condition (2') in Proposition 4.2.13 holds, in particular for $t = 1$.

$$\|\sigma_1(x_m)(a_{x_m}(1)) - a_z\| \to 0 \ . \tag{4.19}$$

By Axiom (γ) in 4.2.8, we have

$$\|\sigma_2(x_m)(e_{1,1}\otimes a_z) - a_z\| \to 0 \ . \tag{4.20}$$

Therefore $\|\sigma_1(x_m)(a_{x_m}(1)) - \sigma_2(x_m)(e_{1,1}\otimes a_z)\| \to 0$. Recall that $\mathrm{Ad}\, U_m = \sigma_2(x_m)^{-1}\sigma_1(x_m)$. It follows that

$$\|\sigma_2(x_m)\mathrm{Ad}\, U_m(a_{x_m}(1)) - \sigma_2(x_m)(e_{1,1}\otimes a_z)\| \to 0 \ ,$$

i.e.

(4.21) $$\|\mathrm{Ad}\, U_m(a_{x_m}(1)) - e_{1,1}\otimes a_z\| \longrightarrow 0 .$$

Since $(a_x)_{x\in\bar{V}}$ also satisfies Axiom (2) of Definition 4.2.11,

(4.22) $$\sup_{t\in[0,1]} \|a_{x_m}(t) - e_{1,1}\otimes a_z\| \longrightarrow 0 .$$

From (4.21) and (4.22) we get

(4.23) $$\|\mathrm{Ad}\, U_m(e_{1,1}\otimes a_z) - e_{1,1}\otimes a_z\| \longrightarrow 0 .$$

Now we apply Lemma 4.2.7 and it follows from (4.23) that

(4.24) $$\sup_{t\in[0,1]} \|\mathrm{Ad}\, U_m^t(e_{1,1}\otimes a_z) - e_{1,1}\otimes a_z\| \longrightarrow 0 .$$

Once again we use (4.22) and obtain

$$\sup_{t\in[0,1]} \|\mathrm{Ad}\, U_m^t(a_{x_m}(t)) - e_{1,1}\otimes a_z\| \longrightarrow 0 .$$

This is exactly

(4.25) $$\sup_{t\in[0,1]} \|a'_{x_m}(t) - e_{1,1}\otimes a'_z\| \longrightarrow 0 .$$

Thus $\phi_\Gamma(a) \in A_\Gamma^2$ for any $a \in A_\Gamma^1$.

It is easy to see that the map ϕ_Γ is a homomorphism. Interchanging the role of A_Γ^1 and A_Γ^2 in the discussion above, we obtain a homomorphism ϕ'_Γ from A_Γ^2 into A_Γ^1 such that for any $a' \in A_\Gamma^2$ with $a' = (a'_x)_{x\in\bar{V}\cap\Gamma}$, we have $\phi'_\Gamma(a') = a = (a_x)_{x\in\bar{V}\cap\Gamma}$, where $a_x = a'_z$ if $z \in V'\cap\Gamma$, while $a_x(t) = \mathrm{Ad}\, W(x)^t(a'_x(t))$, for $0 \le t \le 1$, if $x \in V\cap\Gamma$. Here $W(x)$ are the unitaries given by $\mathrm{Ad}\, W(x) = \sigma_1(x)^{-1}\sigma_2(x) = (\sigma_2(x)^{-1}\sigma_1(x))^{-1} = (\mathrm{Ad}\, U(x))^{-1}$. Therefore

$W(x) = U(x)^{-1}$ and $W(x)^t = U(x)^{-t}$. Consequently $\phi'_\Gamma = \phi_\Gamma^{-1}$ and ϕ_Γ is an isomorphism.

Now we allow the ends of Γ to be vertices of deg 1. We define ϕ_Γ as before but in addition we stipulate that $\phi_\Gamma(a_x) = a'_x$ for $x \in V_1 \cap \Gamma$. Let y be the edge in E_x, where $\deg x = 1$, lying on Γ. There are two possibilities: (i) $\gamma(y) = (x, x_y)$ and $x_y = z \in V' \cap \Gamma$. Since $\phi_{x,y}(a_x) = f_{x,y}(a_x) = a_z = a'_z$, the global glueing condition at x holds as before; (ii) $x_y \in V \cap \Gamma$, thus $\deg(x_y) = 0$. The local glueing condition holds by the discussion in Case (a). Clearly ϕ_Γ is again an isomorphism from A^1_Γ onto A^2_Γ.

Finally, assume y to be an edge in E_x, for $x \in V \cap \Gamma$, not lying on Γ. Then $\deg y = 0$ by the choice of Γ. Therefore we always have $f_{x,y} = \phi_\Gamma \circ f_{x,y}$ by the definition of ϕ_Γ regardless of $\deg(x)$. Q.E.D.

Theorem 4.2.19. *Up to isomorphism the global fibered product* $C^*(T,A,\sigma)$ *of a distinguished tree* (T,A,σ) *of* C^**-algebras is independent of a particular choice of the set* $\sigma = \{\sigma(x)\}_{x\in V_0}$ *of isomorphisms.*

Proof. Given two distinguished trees of C^*-algebras (T,A,σ^i), $i = 1,2$, we show that $C^*(T,A,\sigma^1)$ is isomorphic to $C^*(T,A,\sigma^2)$.

For any $\mathbb{R}$-subtree T_0 of T, recall the definition of the quotient C^*-algebra $A^i_{T_0}$ of $C^*(T,A,\sigma^i)$ for $i = 1,2$. Fix $x_0 \in V$. Let $\tilde{V}$ be all the $x \in V$ with the property that the closed segment Γ_x connecting x_0 to x satisfies the condition that $A^1_{\Gamma_x}$ and $A^2_{\Gamma_x}$ are isomorphic. Let $T_0 = \bigcup_{x\in\tilde{V}} \Gamma_x$. Clearly T_0 is an $\mathbb{R}$-subtree of T. Since $f_{x,y} = f_{x,y} \circ \phi_{\Gamma_x}$, for $y \in E$ not in Γ, $x \in \tilde{V}$, where ϕ_{Γ_x} is defined as in Lemma 4.2.18. We can combine the ϕ_{Γ_x}, $x \in \tilde{V}$ and get an isomorphism ϕ_{T_0} from $A^1_{T_0}$ onto $A^2_{T_0}$.

We claim that $\tilde{V} = V$; hence $T_0 = T$ by the definition for regular $\mathbb{R}$-trees (Definition 2.3.3). Since $A^i_T \simeq C^*(T,A,\sigma^i)$, the theorem follows. Suppose there is a $x \in V\backslash\tilde{V}$. Let Γ be the closed segment connecting x and x_0. We claim that there are only finitely many distinct Γ_z, $z \in V'$ such that $\Gamma \cap \Gamma_z$ is a nondegenerate closed segment (see Remark 3.2.2). In fact, there is $\delta_z > 0$ for each $z \in V'$ such that $B(z,\delta_z) \cap \bar{V} \subset \Gamma_z$, and there is $\delta_x > 0$ for each $x \in V$ such that $B(x,\delta_x) \cap \bar{V} = \{x\}$. Since $\bar{V} \cap T$ is compact, there are only finitely many $B(x,\delta_x)$ and $B(z,\delta_z)$ which cover Γ.

A closer look shows that actually there are only finitely many distinct Γ_z for $z \in V'$ having nonzero intersection with Γ. For if $\Gamma_{z_n} \cap \Gamma \neq \emptyset$, then there is $x_n \in \Gamma_{z_n} \cap \Gamma \cap V$. Since Γ is compact, $\{x_n\}$ has a cluster point $z \in V'$, so Γ_z intersects infinitely many Γ_{z_n}, contradicting Remark 3.3.2.

Now applying Lemma 4.2.12 and Lemma 4.2.14, we see that there is an isomorphism ϕ_Γ from A^1_Γ onto A^2_Γ. Consequently, $\tilde{V} = V$. Q.E.D.

In view of Theorem 4.2.19, we may talk about the C^*-algebra of a distinguished tree (T,d,V,E) and denote it by $C^*(T,A)$ or even $C^*(T)$ instead of $C^*(T,A,\sigma)$, without arousing much confusion.

Lemma 4.2.20 (Rigidity). *Let* $T^i = (T^i,d^i,V^i,E^i)$, $i = 1,2$, *be two distinguished trees. Suppose that there is a degree-preserving map* τ *which sends bijectively* V^1, $V^{1\prime}$ *onto* V^2, $V^{2\prime}$ *respectively, and* E^1_x *bijectively onto* $E^2_{\tau(x)}$ *for all* $x \in V^1$ *such that* $x_n \rightarrow x$ *in*

$\bar{V}^1$ *iff* $\tau(x_n) \to \tau(x)$ *in* $\bar{V}^1$. *Then* τ *is implemented by a similarity from* T *to* T'.

Proof. By Lemma 2.3.4, the bijection between E^1 and E^2 produces a one-to-one correspondence between the connected components of the complement of V^1 and those of V^2. We may define a bijective map σ from T onto T' with the property that $\sigma|_{\bar{V}^1} = \tau$ and which is an affine map along each of those components preserving "orientation". Since the map τ sends $\bar{V}^1$ homeomorphically onto $\bar{V}^2$, it is easy to check that σ is a homeomorphism. Q.E.D.

Theorem 4.2.21. *Suppose* T^1 *and* T^2 *are distinguished trees. Then* T^1 *and* T^2 *are similar iff the* C^**-algebras* $C^*(T^1)$ *and* $C^*(T^2)$ *are isomorphic. In fact every similarity between two distinguished trees is induced by a* C^**-algebra isomorphism, and vice versa.*

Proof. If T^1 and T^2 are similar, then $C^*(T^1)$ and $C^*(T^2)$ are isomorphic by the construction in Definition 4.2.11.

To see the converse, we first consider $\operatorname{Spec}(C^*(T))$ for a distinguished tree T. As a set

$$\operatorname{Spec}(C^*(T)) = \left(\bigcup_{x\in\bar{V}} \operatorname{Spec} A_x\right)/\sim , \tag{4.26}$$

where $p \sim p'$ for $p \in \operatorname{Spec} A_x$ and $p' \in \operatorname{Spec} A_{x'}$, if one of the following holds:

(1) $x, x' \in V$ and there is $y \in E_x$, $\gamma(y) = (x,x')$, such that p (resp. p') corresponds to $\operatorname{Spec} A_y$ under the quotient map from A_x (resp. $A_{x'}$) onto A_y.

(2) $x \in V$, $x' \in V'$ and still for some $y \in E_x$, $\gamma(y) = (x,x')$ such that $\mathrm{Spec}\, A_x = \mathrm{Spec}\, A_y = \{p'\}$ is identified with p by the quotient map from A_x onto A_y.

Now let us consider the topology on $\mathrm{Spec}(C^*(T))$. Recall the definition of $C^*(T)$ (Definition 4.2.11). The local glueing condition (1) provides precisely the equivalence relation in (4.26). Condition (3) of vanishing at ∞ decrees that any closed subset, other than the whole space, is contained in some $\bigcup_{x \in K} \mathrm{spec}\, A_x$, for a compact subset K of $\bar{V}$. Finally the global glueing condition (2) implies that if $z \in V'$ with $x_n \in \bar{V}$ and $x_n \to z$, then any sequence $\{P_n\}$, here $P_n \in \mathrm{Spec}\, A_{x_n}$, converges to the one point set $\mathrm{Spec}\, A_z$. Thus the topology on $\mathrm{Spec}(C^*(T))$ is the coarsest topology that is finer than the usual quotient topology on the quotient space (4.26) and satisfies the two conditions above.

Let V_S be the union of all the points in all $\mathrm{Spec}\, A_y$, for $y \in E$, and all $\mathrm{Spec}\, A_z$, for $z \in V'$. If we let
$HS = \{p \in \mathrm{Spec}(C^*(T)) \mid$ for any $p' \in \mathrm{Spec}(C^*(T))$ there are disjoint neighborhoods of p and p' respectively$\}$. Then $V_S = (\overset{\circ}{HS})^C$. (These are corresponding to the set of separatrices if $T = T(\mathcal{F})$.) Moreover, any component J of the complement of V_S in $\mathrm{Spec}(C^*(T))$ is homeomorphic to an open interval. We may write $J = (J_1, J_2)$, where J_1 and J_2 stand for the two "ends" of J. Fix two sequences $\{P_n^j\}$ in J converging to the two ends J_1 and J_2. Let $\{P_n^j\}'$ be the set of limit points of $\{P_n^j\}$ in $\mathrm{Spec}(C^*(T))$ and let $n_i = \mathrm{Card}\{P_n^j\}'$, $i = 1,2$. The following are the possibilities:

(a) Both $n_1, n_2 \geq 2$. Then J is the interior of some $\mathrm{Spec}(A_{x_1} \underset{A_y}{\vee} A_{x_2})$, the corresponding vertices $x_1, x_2 \in V_0$, with

$\gamma(y) = (x_1,x_2)$, $\gamma(\bar{y}) = (x_2,x_1)$, both y, $\bar{y}$ are of deg 1.

(b) One of $n_1, n_2 < 2$. Then J is the interior of some Spec A_x. If one of n_1, n_2 is 0, then deg $x = 1$. Otherwise deg $x = 0$.

Thus we can recover the graded set V of vertices of T from $\mathrm{Spec}(C^*(T))$. The set V' can also be recovered as the limit set of V_S (i.e. $p \in V'$ iff p is a limit point of some sequence $\{p_n\}$ in V_S, $p_n \neq p$).

One can also reconstruct E_x from $\mathrm{Spec}(C^*(T))$, for every $x \in V$. In fact, suppose that $x_1, x_2 \in V$ are two vertices recovered as in (a). We put $\bar{E}_{x_1} = \{P_n^1\}' \cup \{y\}$ and $\bar{E}_{x_2} = \{P_n^2\}' \cup \{\bar{y}\}$. Clearly the set $\bar{E}_{x_1}$ can be identified with the set E_{x_1} of edges with deg $y = 1$ and deg $y_i = 0$ for all $y_i \in \{P_n^1\}'$. So can $\bar{E}_{x_2}$. Suppose x corresponds to some J of type (b). If $\{P_n^1\}' = \emptyset$, then deg $x = 1$, and the set $E_x = \{P_n^2\}'$. Of course every element $y_i \in E_x$ has deg 0. If $\{P_n^1\}'$ contains one element y, then again $E_x = \{P_n^2\}' \cup \{y\}$ with deg $y = 1$.

Now assume that T^1 and T^2 are distinguished trees and that they have isomorphic C^*-algebras. An isomorphism from $C^*(T^2)$ onto $C^*(T^1)$ induces a homeomorphism V from $\mathrm{Spec}(C^*(T^1))$ onto $\mathrm{Spec}(C^*(T^2))$. By the preceding discussion, V induces a degree-preserving bijection τ mapping $(V^1 \cup V^{1\prime}, E^1)$ onto $(V^2 \cup V^{2\prime}, E^2)$ of the kind described in Lemma 4.2.19 and so by Lemma 4.2.19 τ can be extended to a similarity from (T^1,d^1,V^1,E^1) onto (T^2,d^2,V^2,E^2). Q.E.D.

4.3. C^*-algebras of foliations (Ω,F)

Theorem 4.3.1. *Let* $T = T(F)$ *be the distinguished tree of a foliation* (Ω,F) *with* T_2*-separatrices. Then the global fibered product* $C^*(T,A)$ *of* (T,A) *(Definitions 4.2.9 and 4.2.11) is isomorphic to* $C^*(\Omega,F)$.

Proof. We need to show first that (Ω,F) gives rise to a tree of C^*-algebras (T,A,σ) as defined in Definition 4.2.9 and then that the global fibered product $C^*(T,A,\sigma)$ (Definition 4.2.11) is isomorphic to $C^*(\Omega,F)$.

In 4.2.8 we discussed much of the first part. We have attached C^*-algebras A_x to $x \in V$, and $A_y = K$ to $y \in E$, and also $A_z = K$ to $z \in V'$. Clearly $C^*(L_y$, one leaf$) \simeq C^*(L_z$, one leaf$) \simeq K$ for the leaves L_y and L_z corresponding to $y \in E$ and $z \in V'$. We also have associated the foliated subregion F_x to x and shown that $C^*(F_x,F) \simeq A_x$. Recall from the proof of Theorem 4.1.2 the construction of the isomorphism ϕ_x from $C^*(F_x,F)$ to A_x. Suppose that $\deg x = 0$ and that F_x is conjugate to the region shown in Fig. 4.8. Then there are $(n-1)$ transversals such that for each $t \in [0,1]$, the isomorphism $L^2(\mathbb{R}) \simeq \bigoplus_{i=1}^{n} L^2(U_i(t))$ induces the

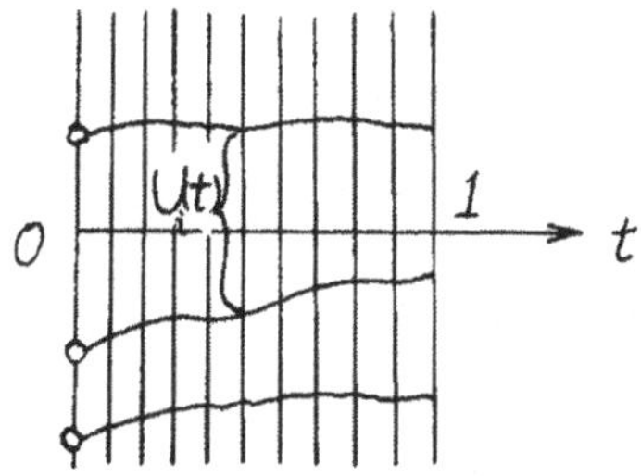

Fig. 4.8

isomorphism $\pi_t(C^*(F_x,F)) \simeq M_n(K)$. We shall denote by $P_i^x(t)$ or $P_i(t)$ the projection from $L^2(\mathbb{R})$ onto $L^2(U_i(t))$. In 4.2.8 we have also shown that for each such set $\{\phi_x\}_{x\in V}$ of isomorphisms, one obtains a set $\{\sigma(x)\}_{x\in V_0}$ of isomorphisms from $M_{n_x}(K)$ to K, where $n_x = \mathrm{Inc}(x) - 1$ such that if we simply define $f_{x,y}$ to be the boundary evaluation map on A_x, then the diagrams (4.1) and (4.2) commute. Thus the surjective homomorphisms $\{\phi_{x,y}\}$ are defined in (3), 4.2.8 accordingly. We have also shown that by choosing the intervals $U_1^x(1)$ appropriately (Fig. 4.3), the set $\sigma = \{\sigma(x)\}$ satisfies the axioms (β) and (γ) in (3), 4.2.8. Axiom (α) is obvious. Therefore to any foliation (Ω,F) we have a tree of C^*-algebras (T,A,σ).

Now we show that there is an isomorphism Ψ from the C^*-algebra $C^*(\Omega,F)$ of a foliation onto the global fibered product $C^*(T,A,\sigma)$. First we verify that the dense subalgebra $C_c(\Omega,F)$ of $C^*(\Omega,F)$ can be imbedded into $C^*(T,A,\sigma)$ isometrically. For each $x \in V$, let π_x be the composition of the quotient map q_x from $C^*(\Omega,F)$ onto $C^*(\Omega_x,F)$ (Proposition 1.5, Part I) and the isomorphism ϕ_x from $C^*(\Omega_x,F)$ onto A_x. For each $z \in V'$, let π_z be the composition of the quotient map q_z and the canonical isomorphism ϕ_z from $C^*(L_z,F)$ onto K (diagrams (4.1) and (4.2) are commuting if $L_z = L_y$ for some $y \in E_x$). Then $(\pi_x(f)) \in \prod_{x\in\bar{V}} A_x$ for any $f \in C_c(G(\Omega,F))$ clearly satisfies the local glueing condition (1) and the condition of vanishing at ∞, (3) in Definition 4.2.11.

Let $z \in V'$ and let $\{x_m\} \subset V$ be a sequence converging to z. Assume that Γ_z is a segment in $T(F)$ containing all x_m and z (Remark 3.3.2) and that all $x_m \in V_0$ and are on one side of z, say $\delta(z)$, without loss of generality. Let Γ be an arc which is a transversal in Ω intersecting L_z and all F_{x_m}. Such Γ can be chosen corresponding to Γ_z and exists by the assumptions on Γ_z. For simplicity, one can imagine that all the leaves intersecting with Γ, as well as all the separatrices bounding Ω_{x_m} have been "straightened up" and if we denote by Ω_Γ those leaves which lie on the one side of L_z, then Ω_Γ is a conjugate foliated submanifold (not saturated, with boundary) of $\mathbb{R}\times[0,1]$ which is foliated with horizontal constant leaves (Fig. 4.9) such that

(1) Γ is just the transversal $(0,S)$, $0 \leq S \leq 1$.

(2) L_z is $\mathbb{R}\times\{1\}$.

(3) Every Ω_{x_m} is contained in $\mathbb{R}\times[S_m, S_m+\delta_m]$, where $S_1 = 0$, $S_1 < S_1+\delta_1 \leq S_2 < S_2+\delta_2 \leq \cdots < 1$ and $S_m \to 1$. In Fig. 4.9 we illustrate for clarity the case $S_m+\delta_m = S_{m+1}$, i.e., the canonical regions Ω_{x_m} and $\Omega_{x_{m+1}}$ are all adjacent.

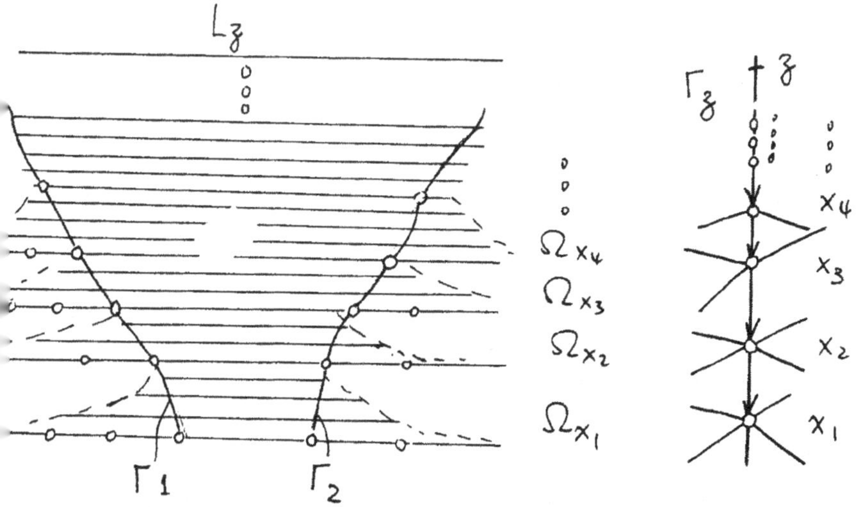

Fig. 4.9

Choose two transversals Γ_1 and Γ_2 such that if the open region between Γ_1 and Γ_2 is denoted by W, then the region $W \cap F_{x_m}$ is just the union of all $U_1^{x_m}(S)$, for $S_m \leq S \leq S_m+\delta_m$. Here the intervals $U_1^{x_m}(S)$ are defined as before, except that the transversal coordinate now is S instead of t and has been rescaled. We write $U_{m,1}(S)$ for $U_1^{x_m}(S)$, and the corresponding projection $P_{m,1}(S)$. Now A_{x_m} has been identified with a C^*-algebra of continuous fields of operators over $[S_m, S_m+\delta_m]$. When $m \to \infty$ and $S \to 1$, the intervals $U_1^{x_m}(S)$ increase to $\mathbb{R}$, and the corresponding projection $P_1^{x_m}(S)$ $*$-strongly converges to I in $\mathcal{L}(L^2(\mathbb{R}))$.

For any $f \in C_c(G(\Omega,F))$, there is some $M > 0$ such that when $m > M$, $\operatorname{supp} f(S,\cdot,\cdot) \subset U_{m,2}(S) \times U_{m,1}(S)$. Recall the isomorphisms $K = K(L^2(\mathbb{R})) \simeq K(\bigoplus_{i=1}^{m} L^2(U_{m,i}(t))) \simeq M_n(K)$ which were used in the construction of the isomorphism ϕ_x from $C^*(F_{x_m},F)$ to A_{x_m}. We always have $P_{m,i}(S)\pi_{x_m}(f)P_{m,i}(S) \in K(L^2(U_{m,i}(S)))$. When $m > M$, then

$$(4.27) \qquad e_{1,1} \otimes (P_{m,1}\pi_{x_m}(f)P_{m,1})(S) = \pi_{x_m}(f)(S) \in M_{n_m}(K) ,$$

for all $S \in [S_m, S_m+\delta_m]$.

Since the open submanifold W of $\mathbb{R} \times [0,1]$ is trivially foliated, there is an obvious isomorphism $C^*(W,F) \simeq C[0,1] \otimes K$. Without loss of generality, we may assume $f \in C_c(G(W,F))$ by throwing away finitely many x_m and cutting off an appropriate chunk of domain(f). Let λ_S be the regular representation of $C^*(\Omega,F)$ at a point $(0,S) \in \Gamma$, $0 \leq S \leq 1$. Then $\lambda_S(f) = (P_{m,1}\pi_{x_m}(f)P_{m,1})(S)$, for $S \in [S_m, S_m+\delta_m]$. We consider $f \in C_c(G(W,F)) \subset C^*(W,F) \subset C^*(\Omega,F)$.

Since $\lambda_S(f) \to \lambda_1(f)$ in norm, as $S \to 1$ and $\lambda_1(f) = \pi_z(f) \equiv f_z$, we have

$$(4.28) \qquad \sup_{t \in [S_m, S_m+\delta_m]} \|(P_{m,1}\pi_{x_m}(f)P_{m,2})(S) - f_z\| \to 0$$

as $m \to \infty$. From (4.27) and (4.28) we obtain

$$(4.29) \qquad \sup_{t \in [S_m, S_m+\delta_m]} \|\pi_{x_m}(f)(S) - e_{1,1} \otimes f_z\| \to 0, \quad m \to \infty.$$

Thus the global glueing condition (2) of Definition 4.2.11 holds for $C_c(\Omega, F)$. It is clear that the imbedding of $C_c(\Omega, F)$ into $C^*(T,A,\sigma)$ is isometric and thus induces an imbedding of $C^*(\Omega, F)$. We have shown in the proof of Theorem 4.2.21 that $\mathrm{Spec}(C^*(\Omega, F))$ is the same as that of $C^*(T,A,\sigma)$. Now applying a C^*-algebra version of the Stone-Weierstrass Theorem (cf. 11.1.1 and 11.1.4 [D]) for type I C^*-algebras, as we did in the proof of Theorem 4.1.2, we see that the inclusion of $C^*(\Omega, F)$ in $C^*(T,A,\sigma)$ is onto. Q.E.D.

Theorem 4.3.2. *The distinguished trees are complete invariants for C*-algebras of foliations with T_2-separatrices. More precisely, the isomorphism classes of C*-algebras of foliations with T_2-separatrices are in exactly one-to-one correspondence with the similarity classes of distinguished trees.*

Proof. By Theorem 3.5.11 every distinguished tree T is similar to an associated tree $T(F)$ for some foliation (Ω, F) with T_2-separatrices. Conversely, the associated tree $T(F)$ for any foliation (Ω, F) with T_2-separatrices is a distinguished tree (Theorem 3.3.4). By Theorem 4.3.1, the C^*-algebra $C^*(\Omega, F)$ is isomorphic to $C^*(T(F))$

for any foliation (Ω, F) with T_2-separatrices. Thus the isomorphism classes of C^*-algebras of foliations with T_2-separatrices are exactly the isomorphism classes of global fibered products of distinguished trees of C^*-algebras. Finally, Theorem 4.2.21 shows that two distinguished trees T and T' are similar if and only if the two corresponding fibered products $C^*(T)$ and $C^*(T')$ are isomorphic. Q.E.D.

Example 4.3.3. Let us consider the two simple examples shown in (a), (b), Fig. 0.1. Their Kaplan diagrams were given in (a), (b) Fig. 1.7, which are clearly non-isomorphic. Thus the two foliations (Ω, F_a) and (Ω, F_b) are not topologically conjugate. It is easy to see that both their associated trees are similar to the one shown in Fig. 3.12(c) which is also the tree associated to the foliation given by Fig. 3.13(c3). Therefore both the C^*-algebras $C^*(\Omega, F_a)$ and $C^*(\Omega, F_b)$ are isomorphic to $C^*(T) \simeq A_1 \vee A_2 \vee A_3 \underset{y_1, y_2}{\vee} (A_1, A_1)$

$$= \left\{(f_1, f_2, f_3, f_4, f_5) \in C_0\left([0,1),\ K \oplus M_2(K) \oplus M_3(K) \oplus K \oplus K\right) \,\middle|\, f_2(0) = \begin{pmatrix} f_1(0) & 0 \\ 0 & T \end{pmatrix},\ f_3(0) = \begin{pmatrix} T & 0 & 0 \\ 0 & f_4(0) & 0 \\ 0 & 0 & f_5(0) \end{pmatrix},\ T \in K\right\}.$$

Example 4.3.4. The trees associated to the foliations given in Fig. 1.8 and Fig. 1.9 are given in 3.2.21 and 3.2.23 (Figs. 3.9 and 3.10) and we saw that they are similar. The C^*-algebras of foliations are isomorphic to

$$\begin{aligned} C^*(T) &= A_\infty \vee (A_1, A_1, \ldots) \\ &= \left\{(f, f_1, f_2, \ldots) \in C_0\left([0,1), (K \otimes K) \bigoplus_{n=1}^{\infty} K\right) \,\middle|\, f(0) = \operatorname{diag}(f_1(0), f_2(0), \ldots)\right\}. \end{aligned}$$

Example 4.3.5. Consider the foliation (Ω,F) given by Fig. 1.3(b), the C^*-algebra $C^*(\Omega,F) \simeq A_2 \vee (A_1,A_2)$. Each of the two points in $\mathrm{Spec}(C^*(\Omega,F))$ corresponding to the two separatrices is closed, but of course not closed in Ω/F with the coarse leaf topology (cf. 1.5.1).

Corollary 4.3.6. *Suppose* (Ω,F) *and* (Ω,F') *are two foliations with* T_2*-separatrices. Then* $C^*(\Omega,F) \simeq C^*(\Omega,F')$ *iff their spectra are homeomorphic.*

Proof. Necessity is clear. Conversely, if the two spectra are homeomorphic, then from the proof of Theorem 4.2.21, the two associated trees are similar. Thus the two C^*-algebras of foliations are isomorphic by Theorem 4.2.21 and Theorem 4.3.1. Q.E.D.

Corollary 4.3.7. *Given an arbitrary foliation* (Ω,F) *of the plane, in the extension (4.0) the quotient* $C^*(S,F)$ *is a continuous trace iff the foliation* (Ω,T) *is of* T_2*-separatrices.*

Proof. Note first that $\mathrm{Spec}(C^*(S,F))$ is homeomorphic to the subset $S \subset \Omega/F$ with the quotient topology. Therefore that $C^*(S,F)$ is a continuous trace implies S is Hausdorff. Conversely, suppose that (Ω,F) is of T_2-separatrices. By Theorem 4.3.1, the C^*-algebra $C^*(\Omega,F)$ is isomorphic to $C^*(T(F))$.

For any $a = (a_x)_{x\in\bar{V}} \in C^*(T(F))$, we let $\pi(a) = (b_y)_{y\in E\cup V'}$, where $b_y = a_y = a_x$ if $y = x \in V'$, and $b_y = \phi_{x,y}(a_x)$ if $y \in E_x$ and $x \in V$. The local glueing condition (1), Definition 4.2.11 implies that $\pi(a)$ is well-defined. Recall the Kaplan transformation Δ from Ω/F onto the chordal system C (1.2.4) and the imbedding ϕ of C into the tree $T(F)$. The set $\phi\circ\Delta(S)$ is closed in $T(F)$.

The condition of vanishing at ∞ and the global glueing condition imply that the isomorphism from $C^*(\Omega,F)$ to $C^*(T(F))$ induces an isomorphism from $C^*(S,F)$ onto $C_0(\phi\circ\Delta(S))\otimes K$. Thus $C^*(S,F)$ is a continuous trace C^*-algebra. Q.E.D.

Bibliography

[A-M] R. C. Alperin and K. Moss, Real-valued Archimedean length functions in groups, to appear (see *Amer. Math. Soc. Abstract* 802-20-54, February, 1983).

[Ar] W. Arveson, The harmonic analysis of automorphism groups, *Proc. Symp. Pure Math.* **38** (1982) 199-260.

[B-D] P. Baum and R. Douglas, K-homology and index theory, *Proc. Symp. Pure Math.* **38** (1982) 117-174.

[Bec] A. Beck, Continuous flows in the plane, *Die Grund. der math. Wissen in Einzel*, Band 201, Springer-Verlag, 1974.

[Ben] I. Bendixon, Sur les courbes défines par des équations différentielles, *Acta Math.* 24 (1901) 1-88.

[B] A. Borel, On the development of Lie group theory, *Nieuw Arch. Wisk.* (3) **27** (1979) 13-25.

[B-G-R] L. Brown, P. Green and M. Rieffel, Stable isomorphism and strong Morita equivalence of C*-algebras, *Pacific J. Math.* **71** (1977) 349-363.

[C1] A. Connes, Sur la théorie non commutative de l'integration, *Lecture Notes in Mathematics* **725** (1979) 19-143, Springer Verlag, New York.

[C2] A. Connes, An analogue of the Thom isomorphism for crossed products of a C*-algebra by an action of $\mathbb{R}$, *Adv. Math.* **39** (1981) 31-55.

[C3] A. Connes, A survey of foliations and operator algebras, *Proc. Symp. Pure Math.* **38** (1982) 521-628.

[C4] A. Connes, Noncommutative differential geometry, Ch. I, II, preprint, IHES, 1983.

[D] J. Dieudonné, Schur functions and group representations, *Astérisque* **87-88** (1981) 7-19.

[Di] J. Diximier, *C*-algebras*, North-Holland, 1917.

[E-H] E. G. Effros and F. Hahn, Locally compact transformation groups and C*-algebras, *Mem. Amer. Math. Soc.* **75** (1967).

[F] T. Fack, Quelques remarques sur le spectre des C*-algebres de feuilletages, preprint, 1984.

[F-S] T. Fack and G. Skandalis, Sur les représentations et idéaux de la C^*-algèbre d'un feuilletage, *J. Oper. Theory* **8** (1983) 95-129.

[Gt] E. C. Gootman, The type of some C^*- and W^*-algebras associated with transformation groups, *Pacific J. Math.* **48**:1 (1973) 98-106.

[G] C. Gutierrez, Smoothing continuous flows and the converse of Denjoy-Schwartz theorem, *Anais Da. Acad. Brasil de Ciên* 51:4 (1979) 581-589.

[H-S] M. Hilsum and G. Skandalis, Stabilité des C^*-algèbres de feuilletages, *Ann. Inst. Fourier* **33** (1983) 201-208.

[Hir-S] M. Hirsch and S. Smale, *Differential Equations, Dynamical Systems, and Linear Algebras*, Academic Press, New York, 1974.

[Kam] E. Kamke, Über die partielle Differentialgleichungen $fz_x + gz_y = h$, *Math. Zeit.* **41** (1936) 55-66; 42 (1936) 287-300.

[K1] W. Kaplan, Regular curve families filling the plane. I, *Duke Math. J.* **7** (1940) 154-185.

[K2] W. Kaplan, Regular curve families filling the plane, II, *Duke Math. J.* **8** (1941) 11-46.

[K3] W. Kaplan, Topology of level curves of harmonic functions, *Trans. Amer. Math. Soc.* **63** (1948) 514-522.

[L] B. Lawson, Foliations, *Bull. Amer. Math. Soc.* **80** (1974) 369-418.

[M] L. Markus, Global structures of ordinary differential equations in the plane, *Trans. Amer. Math. Soc.* **76** (1954) 127-148.

[M-Sc] C. C. Moore and C. Schochet, Foliated spaces, to appear.

[M-S1] J. W. Morgan and P. B. Shalen, Valuations, trees, and degenerations of hyperbolic structures. I, *Ann. Math.* **120** (1984) 401-476.

[M-S2] J. W. Morgan and P. B. Shalen, Valuations, trees, and degenerations of hyperbolic structures. II, to appear in *Ann. Math.*

[N1] T. Natsume, Topological K-theory for codimension one foliations without holonomy, *Adv. Studies in Pure Math.* **5**, Foliations.

[N2] T. Natsume, The C*-algebras of codimension one foliations without holonomy, *Math. Scand.*

[N-T1] T. Natsume and H. Takai, Connes algebras associated to Reeb foliations, preprint, June 1981.

[N-T2] T. Natsume and H. Takai, Connes algebras associated to foliated bundles, preprint, December 1981.

[Nov] S. P. Novikov, The topology of foliations (Russian), *Trudy Moskow, Mat. Obsc.* **14** (1965) 248-278.

[Ped] G. Pedersen, *C*-algebras and Their Automorphism Groups*, Academic Press, New York, 1979.

[Pen] M. Penington, *K-theory and C*-algebras of Lie groups and foliations*, D. Phil. Thesis, Oxford, 1983.

[Po] H. Poincaré, Les courbes définies par des équations différentielles, *J. Math. pures et appliquées* **7**:3 (1881) 375-422; **8**:3 (1882) 251-296; **1**:4 (1885) 167-244; **2**:4 (1886) 151-157.

[Ren] J. Renault, A groupoid approach to C*-algebras, *Lecture Notes in Mathematics* **793** (1980), Springer-Verlag, New York.

[Rie] M. Rieffel, Morita equivalence for C*-W*-algebras, *J. Pure Appl. Alg.* 5 (1974).

[S] J-P. Serre, *Trees*, Springer-Verlag, New York, 1980.

[T] H. Takai, C*-algebras of Anosov foliations, preprint, 1983.

[Tor] A. M. Torpe, K-theory for the leaf space of foliations by Reed components, *J. Func. Anal.* (1985).

[W1] X. Wang, *On the C*-algebras of a class of solvable groups*, Ph.D. Thesis, Part I, preprint.

[W2] X. Wang, On the classification of C*-algebras of Morse-Smale stable flows on closed two manifolds, in preparation.

[Wh] H. Whiteney, Regular families of curves, *Ann. Math.* **34** (1933) 244-270.

[Wil] D. Williams, The topology on the primitive ideal space of transformation group C*-algebras and C.C.R. transformation group C*-algebras, *Trans. Amer. Math. Soc.* **266**:2 (1981).

NOTATIONS AND SYMBOLS

SUBJECT INDEX